I0787947

Thin Film Transistors 15 (TFT) 15

Editor:

Y. Kuo

Sponsoring Divisions:

 Electronics and Photonics

The Japan Society of Applied Physics

Published by
The Electrochemical Society
65 South Main Street, Building D
Pennington, NJ 08534-2839, USA

tel   609 737 1902
fax 609 737 2743
www.electrochem.org

ecstransactions ™

Vol. 98, No. 7

Copyright 2020 by The Electrochemical Society.
All rights reserved.

This book has been registered with Copyright Clearance Center.
For further information, please contact the Copyright Clearance Center,
Salem, Massachusetts.

Published by:

The Electrochemical Society
65 South Main Street
Pennington, New Jersey 08534-2839, USA

Telephone 609.737.1902
Fax 609.737.2743
e-mail: ecs@electrochem.org
Web: www.electrochem.org

ISSN 1938-6737 (online)
ISSN 1938-5862 (print)

ISBN 978-1-60768-902-7 (PDF)

Printed in the United States of America.

Preface

The papers included in this issue of *ECS Transactions* were originally scheduled to be presented in the symposium "Thin Film Transistor Technologies (TFT 15)," held during PRiME 2020 in Honolulu, Hawaii, from October 4-9, 2020. Due to the COVID-19 pandemic, the meeting was changed to an online format. This symposium was sponsored by the Electronics and Photonics Division (EPD) of The Electrochemical Society. This is the 30th year of the symposium, which continuously holds the record of the longest continuous held TFT conference in the world.

The success of this TFT symposium series is greatly contributed by global TFT experts who share their experience and wisdom through presenting technical speeches, authoring papers in special issues of the *ECS Transactions* volumes, co-organizing the meetings, chairing topical sections, and joining panel discussions. The participation of young researchers and students in these symposia also has non-negligible impacts to the growth of this meeting as well as the TFT related fields. The dedicated work of the ECS staff in the past three decades deserves our applause.

There are 40 papers presented in the TFT 15 symposium, divided into eight sessions. Presenters and authors are from universities, industry, and research institutes located in Belgium, France, Ireland, Japan, Korea, Norway, Taiwan, the United Kingdom, and the United States.
1. TFT Device Characteristics and Reliability I
2. TFT Device Characteristics and Reliability II
3. Processes I
4. Processes II
5. Processes III
6. Non-Si and Non-Oxide TFTs
7. TFT Applications in Displays, ICs, and Beyond
8. Posters.

The 21 papers in this *Transactions* volume are divided into the following four chapters. All papers are published as received without alteration of their technical contents.
1. Device Characterization and Reliability
2. Processes
3. Non-Si and Non-Oxide TFTs
4. Applications in Displays, ICs, and Beyond

The progress of the TFT technology in the past 30 years shows the following trends.
- Si based TFTs dominate the production with main research focus on poly-Si TFTs, especially the low cost, high through, highly reliable manufacture technology.
- Oxide TFTs are emerging with targets on simple, reliable, large-area production processes.
- Organic, 2D, and perovskite materials are being actively explored on TFT devices.
- Applications beyond displays, as well as flexible electronics, are hot research topics.

Yue Kuo
Thin Film Nano & Microelectronics Research Laboratory
Texas A&M University, College Station, TX, USA

Thin Film Transistor Technologies 15 Symposium Co-Organizers
Y. Kuo (Texas A&M University)
O. Bonnaud (Université de Rennes I)
A. Flewitt (Cambridge University)
J. Jang (Kyung Hee University)
M. Furuta (Kochi University of Technology)
H. Hamada (Kinki University)
P. Liu (National Chiao Tong University)
A. Nathan (Cambridge University)

Session Chairs and Co-Chairs
O. Bonnaud (Université de Rennes I)
M. Furuta (Kochi University of Technology)
Y. Hirose (Tokyo University) @
K. D. Hirschman (Rochester Institute of Technology)
S. Jeon (Korea Advanced Institute of Science and Technology)
C. Jiang (Cambridge University)
Y. Kuo (Texas A&M University)
P.-T. Liu (National Chiao Tong University)
T. Mohammed-Brahim (Université de Rennes I)
M. Shur (Rensselaer Polytechnic Institute)
Y. Uraoka (Nara Advanced Institute of Science and Technology)
K. S. Chang-Liao (National Tsing Hua University)
H. Hamada (Kinki University)

ECS Transactions, Volume 98, Issue 7
Thin Film Transistors 15 (TFT) 15

Table of Contents

Preface   *iii*

Chapter 2
H03 – Device Characterization and Reliability

(Invited) Hot Carrier Phenomena in Oxide Thin-Film Transistors   3
  J. Tanaka, M. N. Fujii, R. Miyanaga, T. Takahashi, Y. Uraoka, K. Takechi,
H. Tanabe

Characteristics and Applications of CAAC-IGZO FET with Gate Length of 13nm   13
  A. Suzuki, Y. Yuichi, S. Mizukami, K. Tsuda, M. Ito, K. Ohshima, N. Matsumoto,
Y. Yakubo, S. Miyata, N. Okuno, H. Kunitake, S. Sasagawa, T. Ikeda, S. Yamazaki

(Invited) Highly Reliable Metal Oxide Thin Film Transistors for Flexible Devices   29
  Y. Uraoka, D. Corsino, J. P. Bermundo, M. N. Fujii, M. Uenuma

(Invited) UV and Gate Stress Induced Defects in Amorphous Indium Gallium Zinc   39
Oxide Thin Film Transistors and Self-Repair
  J. Jiang, Y. Kuo

(Invited) Device Scalability of InGaZnO TFTs for Next-Generation Displays   47
  S. Oh, S. H. Kim, M. Kim, S. M. Yu, Y. Choi, J. S. Park, J. H. Lim

TCAD Simulation of a 3D NAND Memory Utilizing In-Ga-Zn-Oxide: "3D OS   55
NAND" with 4 V Drive, High Endurance and Density
  H. Kunitake, H. Kimura, K. Tsuda, H. Godo, T. Murakawa, H. Sawai, H. Baba,
S. Sasagawa, T. Ikeda, S. Yamazaki

(Invited) Highly Stable Self-Aligned Top-Gate Indium Gallium Zinc Oxide Thin-Film   69
Transistors for High-Resolution OLED TV and Mobile Displays
  J. B. Kim, Y. C. Tsai, R. Lim, Z. Wang, M. Hao, S. W. Wang, J. W. Park, L. Zhao,
M. Bender, D. K. Yim, S. Y. Choi

Chapter 3
H03 – Processes

Interpretation of Donor Activation in Boron and Argon Implanted Self-Aligned Bottom-Gate IGZO TFTs   81
  M. S. Kabir, R. R. Chowdhury, R. G. Manley, K. D. Hirschman

Low-Temperature Processed Metal-Semiconductor Field-Effect Transistor with In-Ga-Zn-O Channel Deposited by Ar+O_2+H_2 Sputtering   89
  Y. Magari, M. Furuta

(Invited) Improved Copper Electrode Integration for Thin Film Electronics on Glass   97
  H. Kim, B. Zhu, M. H. Huang, R. Vaddi, R. G. Manley

Effect of Glass Substrate on the Film Properties of Poly Silicon by Excimer Laser Annealing   109
  B. Zhu, R. Vaddi, M. H. Huang, H. Kim, R. G. Manley

Influence of Glass Surface Modification on Thin Film Copper Electrodes   117
  M. H. Huang, Y. Shi, B. Zhu, R. Vaddi, H. Kim, R. G. Manley

Edge-Directed LTPS Via Flash Lamp Annealing Using a Cr Adhesion Layer for Improved Wettability   131
  G. Packard, A. Rosenfeld, M. Hum, R. G. Manley, K. D. Hirschman

Flash Lamp Annealed LTPS TFTs with ITO Bottom-Gate Structures   141
  G. Packard, A. Rosenfeld, R. G. Manley, K. D. Hirschman

Photoconductive Solution Processed ZnO Quasi-superlattice Films   151
  D. Buckley, S. Inguva, D. McNulty, V. Z. Zubialevich, P. Parbrook, F. Gity, P. K. Hurley, C. O'Dwyer

Chapter 4
H03 – Non-Si and Non-Oxide TFTs

(Invited) Percolation Carbon Nanotube Thin Film Transistors   161
  M. Shur, J. Park, Y. Zhang, X. Liu, T. Ytterdal

(Invited) Optimizing Material Systems for All-Inkjet-Printed Organic Thin-Film Transistors 173
  C. Jiang, A. Nathan

Chapter 5
H03 – Applications in Displays, ICs, and Beyond

(Invited) Display and LSI Applications of Oxide Semiconductor LSIs (OS LSIs) Using Crystalline In-Ga-Zn Oxide (IGZO): Applications Related to Coronavirus COVID-19 Pandemic 185
  T. Onuki, Y. Okamoto, T. Aoki, T. Matsuzaki, M. Kozuma, H. Kunitake, R. Motoyoshi, H. Kimura, Y. Yamane, S. Sasagawa, S. Yamazaki

(Invited) Sub-40mV Sigma V_{TH} Igzo nFETs in 300mm Fab 205
  J. Mitard, L. Kljucar, N. Rassoul, H. F. W. Dekkers, M. van Setten, A. Chasin, G. Pourtois, A. Belmonte, R. Delhougne, G. L. Donadio, L. Goux, M. Nag, C. Wilson, Z. Tokei, J. I. D. A. Borniquel, S. Steudel, G. S. Kar

(Invited) Hafnia Ferroelectric Device for Semiconductor, Sensor, and Display Applications 219
  S. Jeon

P3HT:ZnS Based Photovoltaic Devices with Enhanced Performance Assisted by Oxidised Carbon Nanotubes 225
  C. Wei, M. T. Bishop, Y. Wang, F. Gao, C. Wang, G. Z. Chen

Author Index 233

Facts about ECS

The Electrochemical Society (ECS) is an international, nonprofit, scientific, educational organization advancing the theory and practice of electrochemistry and solid state science and technology, and allied subjects. The Society was founded in Philadelphia in 1902 and incorporated in 1930. There are currently over 8,000 members from around the globe representing 13 technical division and 23 geographical sections and a growing student membership program with over 100 student chapters. The Society is also supported by more than 2,000 corporations, government agencies, and academic institutions through institutional membership, corporate programs, and subscriptions.

The technical activities of the Society are carried on by divisions. Sections of the Society host symposia, programs, and events focused on their respective geographic regions. Major international meetings of the Society are held in the spring and fall of each year. At these meetings, the divisions and partnered organizations hold general sessions and sponsor symposia on specialized subjects.

The Society has an active publications program that includes the following:

Journal of The Electrochemical Society — (JES) is the flagship journal of The Electrochemical Society and the oldest peer-reviewed journal in its field. Since its founding in 1902, JES has evolved into one of the most highly cited and prestigious journals in electrochemistry and materials science with a cited half-life of greater than 10 years.

ECS Journal of Solid State Science and Technology — (JSS) is a peer-reviewed journal covering fundamental and applied areas of solid state science and technology, including experimental and theoretical aspects of the chemistry, and physics of materials and devices.

ECS Transactions (ECST) — is the official conference proceedings publication of The Electrochemistry Society — a high-quality venue for authors and an excellent resource for researchers. ECST offers the full-text content of proceedings from ECS meetings and ECS sponsored conferences.

The Electrochemical Society Interface — *Interface* is an authoritative yet accessible publication for those in the field of solid state and electrochemical science and technology. Published quarterly, this full-color magazine contains technical articles about the latest developments in the field, and presents news and information about the Society.

ECS Books Series — ECS books and monographs provide authoritative, detailed accounts of specific topics in electrochemistry and solid state science and technology. These titles are sponsored by ECS and published in cooperation with noted publishers such as John A. Wiley & Sons.

For more information on these publications and other Society activities, visit the ECS website:

www.electrochem.org

Chapter 2

H03 – Device Characterization and Reliability

Hot Carrier Phenomena in Oxide Thin-Film Transistors

J. Tanaka[a], M. N. Fujii[b], R. Miyanaga[b], T. Takahashi[b],
Y. Uraoka[b], K. Takechi[a], and H. Tanabe[a]

[a]Tianma Japan, Ltd., Kawasaki 211-8666, Japan
[b]Nara Institute of Science and Technology, Ikoma 630-0192, Japan

We observed hot carrier-induced photon emissions in oxide TFTs with structures of BCE-type and TG-type for the first time in addition to ESL-type. In this paper, the effects of these TFT structures on the hot carrier phenomena are described using TFT characteristics, photon-emission analysis, and TCAD simulation. We observed the degradation of TFT characteristics in BCE-type and TG-type, which was similar to that in Si-based devices, along with the photon emission. In contrast, for the ESL-type, we found the behavior depending on the drain–ESL overlap length. When this overlap length was under 1μm, photon emission and no significant shift of TFT characteristics were observed; however, when it was 3μm, no photon emission and a significant shift of TFT characteristics were observed. Based on the potential distributions near the drain obtained by the TCAD simulation, we propose the degradation mechanism of ESL-type depending on the drain–ESL overlap length.

Introduction

Since the original study conducted by Nomura et al., thin-film transistors (TFTs) using amorphous oxide semiconductors, such as a-InGaZnO$_4$ (a-IGZO), have caused the rapid evolution of TFT-based electronics owing to their characteristics of high mobility, small S value, small leakage current, and good uniformity across a large area (1-5). Although amorphous oxide TFTs are mass-produced in some displays, such as liquid crystal displays (LCDs) and organic light emitting diode (OLED) displays recently, their applications are expected to expand further (6-8). Up to now, mainly three types of oxide TFTs have been researched and developed. The first is a back channel etch (BCE)-type TFT that applies a process similar to that of the amorphous silicon (a-Si) TFT and etches source and drain (S/D) metal to form a back channel (6, 9). The second is an etch stop layer (ESL)-type TFT that is protected by an ESL in advance to prevent the back-channel interface from being exposed to etching (7, 10). The third is a self-aligned top-gate (TG)-type TFT exhibiting a structure similar to that of the low-temperature poly-crystalline silicon (LTPS) TFT with low parasitic capacitance (7, 11-13). For practical use, it is important to achieve stable electrical characteristics in each type of oxide TFT. Thus, various studies have been conducted on the instability of the characteristics of oxide TFTs such as the threshold voltage shift (V$_{th}$) and a light-deterioration phenomenon caused by the application of bias voltage or light illumination (14-16). However, in the future, we believe that the degradation phenomenon of oxide TFTs under current stress (under high V$_{gs}$ and high V$_{ds}$) will become more important for the following two reasons:

(a) the development of the application side is progressing, and the number of current-driven devices, such as OLEDs and μLEDs, is expected to increase in the future (7-8, 17); (b) in terms of materials, high-mobility oxide-semiconductor materials have been actively developed (18-20). Therefore, we believe that the hot carrier effect under high V_{ds} in oxide TFTs should be well understood.

Hot carrier-induced degradation is a well-known phenomenon in Si-based devices such as single-crystalline-silicon metal-oxide-semiconductor field-effect transistors (MOS-FETs) and polycrystalline-Si (poly-Si) TFTs by high drain electric fields; it accelerates the decrease in ON currents and increase in V_{th} shifts (21-32). This hot carrier-induced degradation is caused by the local depletion at the drain region in the channel; it corresponds to the formation of the potential barrier due to hot electron injection into the gate insulator and generation of interface trap between the channel and the gate insulator (22). The photon emission from the device is a particularly remarkable phenomenon (21, 23-32). The photon-emission spectrum of the hot carrier is around the visible light region, and the photon emission is understood to originate from bremsstrahlung (21, 23-32). It is also known that the number of emitted photons is maximum at a certain V_{gs} (27).

In relation to oxide TFTs, many researchers have predicted the existence of hot carriers as one of the degradation models of device reliability from transfer characteristics and capacitance–voltage (C–V) characteristics (33-39). Recently, we observed significant V_{th} shift and photon emission at the drain region under high V_{ds} for ESL-type TFTs with a S/D that overlapped (<1μm) with the ESL using an emission microscope; we also observed that the photon energies have a wide range of approximately 1.38–2.48 eV, and these results suggested that this photon emission was caused by hot carriers (10). As per our knowledge, this previous study was the first that observed photon emission from oxide TFTs.

In this study, we investigated the hot carrier effects in BCE-, TG-, and ESL-type TFTs using TFT characteristics, photon-emission analysis, and technology computer-aided design (TCAD) simulation. For the BCE- and TG-type TFTs, the observed degradation was similar to that in Si-based devices, whereas for ESL-type TFTs, the degradation was different depending on the device structure. Finally, we propose the degradation mechanism of ESL-type TFTs based on potential distributions near the drain obtained by TCAD simulation.

Experiments

As shown in Figs 1 (a), (b), and (c), we fabricated BCE-, TG-, and ESL-type TFTs using oxide semiconductor material as an active layer, respectively. For the BCE-type TFT, a gate metal thin film was deposited on a glass substrate via sputtering. A gate insulator was deposited on the gate electrode by plasma-enhanced chemical vapor deposition (PECVD). A 90 nm-thick a-IGZO (In:Ga:Zn = 1:1:1) thin film was deposited on the gate insulator via sputtering at room temperature in Ar/O_2 gas. This IGZO channel was patterned by wet etching. S/D metal films were deposited via sputtering and patterned by dry etching (DE). A passivation layer was deposited on the fabricated device via PECVD. An organic film was coated as a planarization layer.

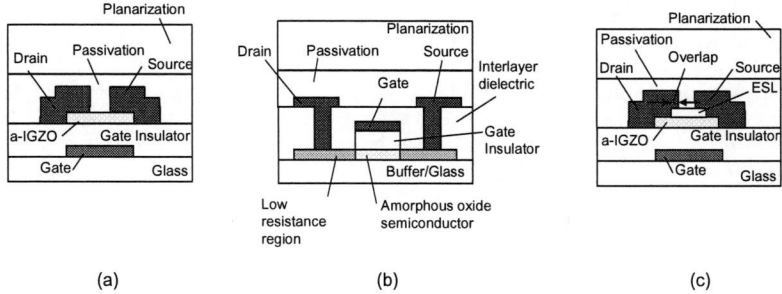

(a) (b) (c)

Figure 1 Cross sections of the (a) BCE-type TFT, (b) TG-type TFT, and (c) ESL-type TFT

The main fabrication process for the TG-type TFT has been described in a previous report (13). In this study, an amorphous oxide semiconductor material with higher mobility than that of a-IGZO was adapted to the active layer. Low-resistance regions in the active layer were self-aligned to the gate via Helium plasma treatment (12).

The main fabrication process for the ESL-type TFT has been described in a previous report (10). Two type of the drain–ESL overlap length were prepared with 3μm and <1μm. In addition, we fabricated an ESL-type TFT with indium tin oxide (ITO) as the gate electrode to observe photon emission from the glass substrate side.

The photon emission was observed by an emission microscope (Hamamatsu Photonics K.K. PHEMOS-200). This system has a photon-counting camera with a GaAs photo cathode. The wavelength-sensitivity range was 500–900 nm from the spectral-sensitivity characteristics. Combining this system with a semiconductor parameter analyzer (Keysight Technologies, Inc. B1500A), the hot carrier-induced degradation can be evaluated by an emission analysis and measurements of electrical properties. Further, by changing the manipulator, the analysis of photon emission from the glass substrate side is possible.

Results and Discussion

We observed hot carrier-induced photon emissions in oxide TFTs with BCE- and TG-type TFTs. In contrast, for the ESL-type, we found the behavior depending on the drain–ESL overlap length. When this overlap length was under 1μm, photon emission was observed. However, when it was 3 μm, photon emission was not observed. Detailed results for each type TFT are presented below.

BCE-type TFT

Figure 2 (a) shows the I_{ds}-V_{gs} characteristics that were measured twice under high V_{ds} (forward characteristics), and then, the S/D bias was reversed and measured twice (reverse characteristics) for the BCE-type TFT. The inset shows the characteristics in log scale. In the reverse characteristic, the sub-threshold region is deteriorated, and the ON current is lower than that in the forward characteristic. Because this degradation is similar to the hot carrier degradation in Si-based devices, it is speculated that the cause of this degradation is the increase in defects near the drain due to stress during the forward-characteristic measurements (32).

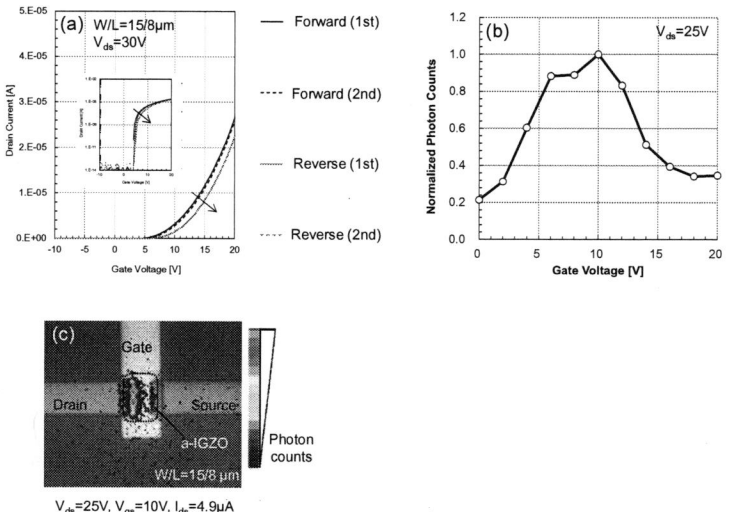

Figure 2 (a) I_{ds}-V_{gs} curves under a high V_{ds}, (b) V_{gs} dependence on the photon counts, and (c) the photon-emission image of the BCE-type TFT

As the result of the emission analysis, Figs 2 (b) and (c) show the V_{gs} dependence of the photon counts and the photon-emission image, respectively. When V_{gs} was increased from 0 V, photon emission began to be observed from $V_{gs} = 4$ V, and the photon counts increased, reaching the maximum at $V_{gs} = 10$ V. When V_{gs} was further increased, the photon counts decreased. This behavior is also observed in Si-based devices and is one of the characteristics of photon emission induced by hot carriers.

Further, as seen in Fig. 2 (c), photon emission was observed at the drain edge, similar to the case of Si-based devices. In this area, the drain electrode opposite to the source electrode and IGZO are in contact with each other. These results reveal that the degradation due to hot carriers in the BCE-type TFT was similar that in Si-based devices. To the best of our knowledge, this is the first time that hot carrier-based photon emission has been observed in BCE-type TFTs. In addition, two positions where most photons were irradiated correspond to those where the drain slightly extends by approximately 0.2 μm to the source. Because the electric field was concentrated at these two positions, photon emissions were estimated to be particularly strong there.

TG-type TFT

Figure 3 (a) shows the I_{ds}-V_{gs} characteristics of the TG-type TFT, measured in the same manner as in Fig. 2(a). The reverse characteristic of this TFT also has a lower ON current than that in the case of the forward characteristic. Further, as shown in Fig. 3 (b), when V_{gs} was increased from 0 V, photon emission started from $V_{gs} = 4$ V, and the photon counts were maximum at $V_{gs} = 10$ V. In addition, as seen in Fig. 3 (c), photon emission was observed at the gate edge opposite to the drain. Because there was no significant change in gate current during these measurements, photon emission was estimated to have originated not from the leakage current of the gate insulator between the gate electrode and the active layer but from the interface between the channel region and the low-resistance region of the active layer. According to these results, the degradation due to hot carriers was observed in the TG-type TFT as well as the BCE-type TFT and Si-based devices. We believe that this is also the first time that photon emission from hot carriers has been observed in TG-type TFTs.

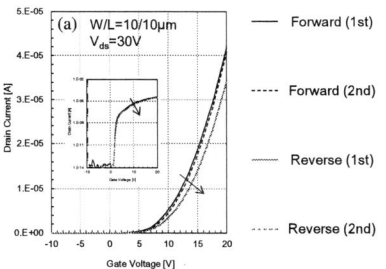
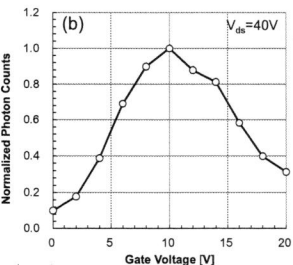

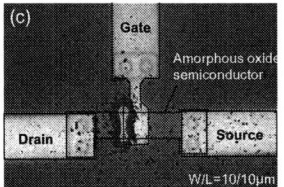

$V_{ds}=40V$, $V_{gs}=10V$, $I_{ds}=19μA$

Figure 3 (a) I_{ds}-V_{gs} curves under a high V_{ds}, (b) V_{gs} dependence on the photon counts, and (c) the photon-emission image for the TG-type TFT

Furthermore, in Fig. 3 (c), two positions that were irradiated with the most photons correspond to edges of the active layer in the channel width direction, where the gate electrode (and the gate insulator) crosses over a step of the active layer. These positions have been reported to be one of the causes of the hump characteristics and instability of oxide TFTs (13, 39). We also reported that the carrier concentration at the edge of the active layer in the channel width direction was higher than that at the center in TG-type TFTs with hump characteristics in the positive gate bias stress (PBS) (13). Although the results in Fig. 3 can be related to the cause of the hump characteristics, further investigation is required to clarify the detailed mechanism.

ESL-type TFT

Figure 4 (a) shows the I_{ds}-V_{gs} characteristics of the ESL-type TFT with the drain–ESL overlap length <1 µm, measured in the same manner as in Fig. 2(a). No significant shift in the reverse characteristic was observed in the I_{ds}-V_{gs} characteristics. Further, as shown in Figs 4 (b) and (c), photon emission was observed near the drain, and the photon counts were maximum at V_{gs} = 12 V, like in the previous study (10). This behavior is almost similar to that of the BCE-type TFT. Next, we evaluated an ESL-type TFT with the drain–ESL overlap length 3µm. As seen in Fig. 5(a), the forward characteristic showed no significant changes, while the reverse characteristic considerably shifted to the positive direction. This result suggests that a potential barrier was formed in the drain region under a high V_{ds} stress and caused local depletion (10). However, photon emission was not observed, as shown in Fig. 5 (b).

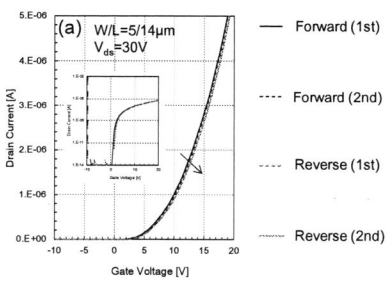
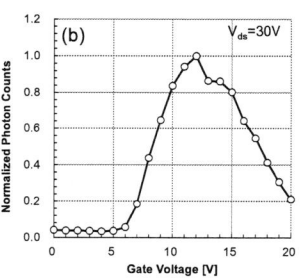
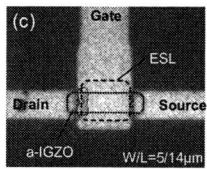

V_{ds}=30V, V_{gs}=12V, I_{ds}=1.0µA

Figure 4 (a) I_{ds}-V_{gs} curves under a high V_{ds}, (b) V_{gs} dependence on the photon counts, and (c) the photon-emission image for the ESL-type TFT (drain–ESL overlap length less than 1 µm)

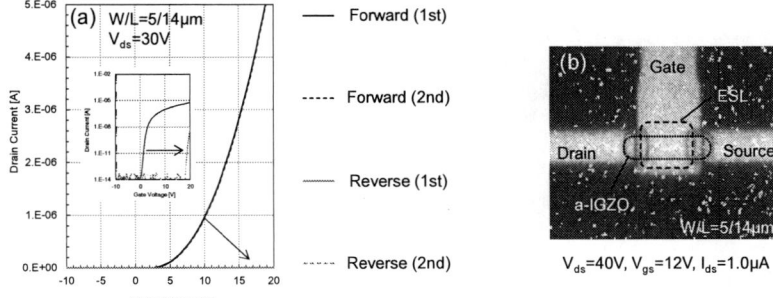

Figure 5 (a) I_{ds}-V_{gs} curves under a high V_{ds} and (b) the photon-emission image for the ESL-type TFT (drain–ESL overlap length = 3 μm)

It was possible that the position where the drain electrode contacts IGZO might be hidden by the drain-electrode overlap when the ESL-type TFT was observed from above, because the BCE-type TFT irradiated photons at the position where the drain electrode contacted IGZO (Fig. 2 (c)). Therefore, a TFT with a gate formed of a transparent conductive-film ITO was fabricated and observed from below (from the glass substrate side). As a result, no photon emission was observed when the drain–ESL overlap length was 1μm or more, whereas photon emission was observed when it was less than 1μm and when observed from above. Thus, in the ESL-type TFT with a drain–ESL overlap length of 3μm, the reason why photon emission was not observed was that not enough hot carriers were generated near the drain. These results suggest that the degradation under high V_{gs} and V_{ds} stress strongly depends on the drain–ESL overlap region for the ESL-type TFT.

In order to obtain insight on degradation depending on the drain–ESL overlap region, we conducted TCAD simulation (ATLAS, by Silvaco Inc.). Figures 6 (a) and (b) present the calculation results obtained using the TCAD simulations of the potential distributions near the drain for the ESL-type TFT under $V_{gs} = 12$ V and $V_{ds} = 30$ V when the drain–ESL overlap lengths were 0μm and 3μm, respectively. On one hand, when the drain–ESL overlap length was 0μm (Fig. 6 (a)), the potential changed steeply in the region where the drain edge contacted IGZO (the region surrounded by the red square). This indicates that the electric field was concentrated. This position corresponds to the position where photon emission was observed in Figs 2 (c) and 4 (c). On the other hand, when the drain–ESL overlap length was 3μm, the electric field in IGZO was relaxed near the drain because the potential changed gradually in the drain–ESL overlap region. However, the potential in the ESL where the drain overlapped was distributed along the thickness direction from the bottom to the top.

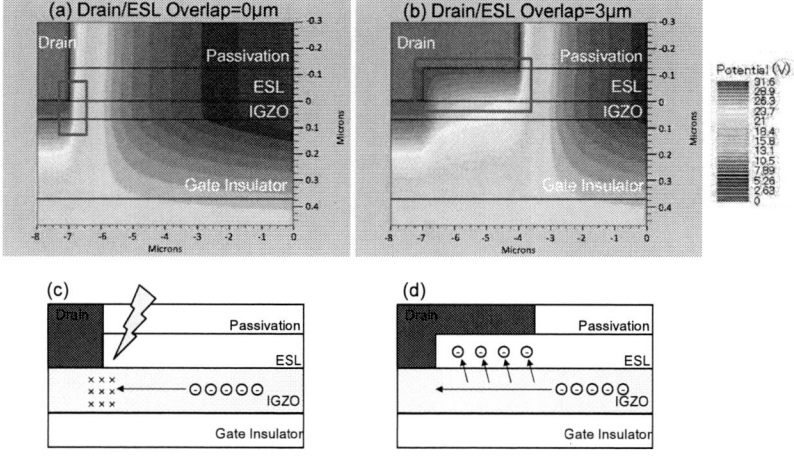

Figure 6 Potential distributions near the drain calculated by the TCAD simulation and the degradation model for the ESL-type TFT when the drain–ESL overlap lengths are (a), (c) 0μm and (b), (d) 3μm

Based on these results, we propose the following degradation mechanism of the ESL-type TFT. On one hand, when the drain–ESL overlap length is 0μm, the electric field concentrated near the drain causes the carriers to obtain a high energy, hot carriers are generated, defects are generated in IGZO, photons are emitted, and the defects formed near the drain degrade the TFT characteristics (40-42). On the other hand, when the drain–ESL overlap length is 3μm, the electric field in IGZO is relaxed in the drain–ESL overlap region, and thus, the generation of hot carriers is less than that shown in Fig. 6 (a). As a result, no photon emission is observed. However, owing to the potential distribution in the ESL film where the drain overlaps, carriers are injected from IGZO into the ESL film and trapped in it. As a result, a local potential barrier that we proposed in a previous study was formed near the drain, and the reverse characteristics significantly shifted to the positive direction (10).

As described earlier, we propose that the difference in degradation depending on the drain–ESL overlap length for the ESL-type TFT is attributed to the difference in potential distribution between the IGZO and ESL films.

Conclusions

We investigated the effects of TFT structures, such as BCE-type, TG-type and ESL-type, on the hot carrier phenomena using TFT characteristics, photon-emission analysis, and TCAD simulation. In this study, we observed photon emissions induced hot carrier in oxide TFTs with BCE-type and TG-type structures for the first time. We also observed the degradation of TFT characteristics in BCE- and TG-type TFTs, which was similar to that in the case of Si-based devices, along with photon emission. In contrast, for the ESL-type TFT, we found the photo emission and the degradation of TFT characteristics depended on the drain–ESL overlap length. When this overlap length was under 1µm, photon emission and no significant shift of TFT characteristics were observed (almost similar to the case of the BCE-type TFT). However, when it was 3µm, no photon emission and a significant shift were observed. Based on the results of the TCAD simulation, we proposed that the differences in photon emission and degradation of TFT characteristics were attributed to the potential distributions near the drain for the ESL-type TFT. These results indicate that the TFT structure is a parameter that affects the hot carrier phenomena, and further research is necessary for highly reliable current-drive TFTs.

References

1. K. Nomura, H. Ohta, A. Takagi, T. Kamiya, M. Hirano, and H. Hosono, *Nature (London)* **432,** 488 (2004).
2. H. D. Kim, J.-S. Park, Y. G. Mo, and S. S. Kim, *Digest of Int. Meeting on Information Display*, 35 (2009).
3. M. Ito, M. Kon, T. Okubo, M. Ishizaki, and N. Sekine, *Proc. IDW/AD'05*, 845 (2005).
4. S.-M. Park, C. Ha, K.-C. Choi, S.-J. Yun, H.-S. Kim, W.-C. Jung, K.-S. Yang, H.-S. Seo, B.-C. Kim, and S.-Y. Cha, *Proc. IDW'11*, 591 (2011).
5. T. Kamiya, K. Nomura, and H. Hosono, *Sci. Technol. Adv. Mater.* **11,** 044305 (2010).
6. T. Matsuo, S. Mori, A. Ban, and A. Imaya, *SID 2014 Digest*, 83 (2014).
7. Y. -M. Ha, S. K. Kim, H. Choi, S. -G. Lee, K. -S. Park, and I. Kang, *SID 2016 Digest,* 943 (2016).
8. T. -K. Chang, C. -W. Lin, and S. Chang, *SID 2019 Digest*, 547 (2019).
9. S. Iwamatsu, K. Takechi, T. Yahagi, Y. Watanabe, H. Tanabe, and S. Kobayashi, *Jpn. J. Appl. Phys.* **48,** 04C091 (2009).
10. T. Takahashi, R. Miyanaga, M. N. Fujii, J. Tanaka, K. Takechi, H. Tanabe, J. P. Bermundo, Y. Ishikawa, and Y. Uraoka, *Appl. Phys. Express* **12,** 094007 (2019).
11. N. Morosawa, Y. Ohshima, M. Morooka, T. Arai, and T. Sasaoka, *SID'11 Digest*, 479 (2011).
12. K. Takechi, Y. Kuwahara, J. Tanaka, and H. Tanabe, *Jpn. J. Appl. Phys.* **58,** 038005 (2019).
13. Y. Kuwahara, K. Takechi, J. Tanaka, and H. Tanabe, *IEEE Electron Device Lett.* **40,** 1273 (2019).
14. K. Takechi, M. Nakata, T. Eguchi, H. Yamaguchi, and S. Kaneko, *Jpn. J. Appl. Phys.* **48,** 010203 (2009).
15. J. Tanaka, Y. Ueoka, K. Yoshitsugu, M. Fujii, Y. Ishikawa, Y. Uraoka, K. Takechi, and H. Tanabe, *ECS Journal of Solid Science and Technology*, 4(7) Q61 (2015).
16. Y. Uraoka, J. P. Bermundo, M. N. Fujii, M. Uenuma, and Y. Ishikawa, *Jpn. J. Appl. Phys.* **58,** 090502 (2019).
17. J. G. Um, D. Y. Jeong, Y. Jung, J. K. Moon, Y. H. Jung, S. Kim, S. H. Kim, J. S. Lee, and J. Jang, *Advanced Electronic Materials* vol. **5**, issue 3, 1800617 (2019).
18. M. Ochi, S. Morita, Y. Takanashi, H. Tao, H. Goto, T. Kugimiya and M. Kanamaru, *SID 2015 Digest*, 853 (2015).
19. E. Fukumoto, T. Arai, N. Morosawa, K. Tokunaga, Y. Terai, T. Fujimori, and T. Sasaoka, *Journal of the SID* **19/12,** 867 (2011).

20. T. Kizu, N. Mitoma, M. Miyanaga, H. Awata, T. Nabatame, and K. Tsukagoshi, *J. Appl. Phys.* **118**, 125702 (2015).

21. S. Tam and C. Hu, *IEEE Trans. Electron Devices* **ED-31**, 1264 (1984).

22. K.-L. Chen, S. A. Saller, I. A. Groves, and D. B. Scott, *IEEE Trans. Electron Devices* **ED-32**, 386 (1985).

23. A. Toriumi, M. Yoshimi, M. Iwase, Y. Akiyama, and K. Taniguchi, *IEEE Trans. Electron Devices* **ED-34**, 1501 (1987).

24. M. Lanzoni, E. Sangiorgi, C. Fiegna, M. Manfredi, and B. Riccó, *IEEE Electron Device Lett.* **12**, 341 (1991).

25. Y. Uraoka, N. Tsutsu, Y. Nakata, and S. Akiyama, *IEEE Trans. Semicond. Manuf.* **4**, 183 (1991).

26. A. L. Lacaita, F. Zappa, S. Bigliardi, and M. Manfredi, *IEEE Trans. Electron Devices* **40**, 577 (1993).

27. F. V. Farmakis, C. A. Dimitriadis, J. Brini, G. Kamarinos, V. K. Gueorguiev, and T. E. Ivanov, *J. Appl. Phys.* **85**, 6917 (1999).

28. Y. Uraoka, T. Hatayama, T. Fuyuki, T. Kawamura, and Y. Tsuchihashi, *Jpn. J. Appl. Phys.* **39**, L1209 (2000).

29. Y. Uraoka, T. Hatayama, T. Fuyuki, T. Kawamura, and Y. Tsuchihashi, *Jpn. J. Appl. Phys.* **40**, 2833 (2001).

30. Y. Uraoka, Y. Morita, H. Yano, T. Hatayama, and T. Fuyuki, *Jpn. J. Appl. Phys.* **40**, 2833 (2002).

31. Y. Uraoka, N. Hirai, H. Yano, T. Hatayama, and T. Fuyuki, *IEEE Electron Device Lett.* **24**, 236 (2003).

32. C. Hu, S. C. Tam, F. -C. Hsu, P. -K. Ko, T. -Y. Chan, and K. W. Terril, *IEEE Transactions on Electron Devices*, vol. **ED-32**, No. 2 (1985).

33. S.-H. Choi and M.-K. Han, *Appl. Phys. Lett.* **100**, 043503 (2012).

33. M.-Y. Tsai, T.-C. Chang, A.-K. Chu, T.-C. Chen, T.-Y. Hsieh, Y.-T. Chen, W.-W. Tsai, W.-J. Chiang, and J.-Y. Yan, *Thin Solid Films* **528**, 57 (2013).

35. T.-Y. Hsieh, T. -C. Chang, Y. –T. Chen, P.-Y. Liao, T. –C. Chen, M. –Y. Tsai, Y. –C. Chen, B. -W. Chen, A. –K. Chu, C. -H. Chou, W. -C. Chung, and J. -F. Chang, *IEEE Electron Device Lett.* **34**, 638 (2013).

36. S. Urakawa, S. Tomai, Y. Ueoka, H. Yamazaki, M. Kasami, K. Yano, D. Wang, M. Furuta, M. Horita, Y. Ishikawa, and Y. Uraoka, *Appl. Phys. Lett.* **102**, 053506 (2013).

37. S. M. Lee, W.-J. Cho, and J. T. Park, *IEEE Trans. Device Mater. Rel.* **14**, 471 (2014).

38. H.-J. Lee, S. H. Cho, K. Abe, M.-J. Lee, and M. Jung, *Sci. Rep.* **7**, 9782 (2017).

39. G.-F. Chen, T. -C. Chang, H. -M. Chen, B. -W. Chen, H. -C. Chen, C. -Y. Li, Y. -H. Tai, Y. -J. Hung, K. -J. Chang, K. -C. Cheng, C. -S. Huang, K. -K. Chen, H.-H. Lu, and Y. -H. Lin, *IEEE Electron Device Lett.* **38**, 334 (2017).

40. T. Takahashi, M. N. Fujii, R. Miyanaga, M. Miyanaga, Y. Ishikawa, and Y. Uraoka, *Appl. Phys. Express* **13,** 054003 (2020).

41. M. Fujii, Y. Ishikawa, M. Horita, and Y. Uraoka, *Appl. Phys. Express* **4,** 104103 (2011).

42. H.-J. Lee, S. H. Cho, K. Abe, M.-J. Lee, and M. Jung, *Sci. Rep.* **7**, 9782 (2017).

Characteristics and Applications of CAAC-IGZO FET with Gate Length of 13 nm

A. Suzuki, Y. Yanagisawa, S. Mizukami, K. Tsuda, M. Ito, K. Ohshima, N. Matsumoto, Y. Yakubo, S. Miyata, N. Okuno, H. Kunitake, S. Sasagawa, T. Ikeda, and S. Yamazaki

Semiconductor Energy Laboratory Co., Ltd. 398 Hase, Atsugi, Kanagawa 243-0036
Japan
Phone: +81-46-248-1131, Fax: +81-46-270-3751

> A field-effect transistor (FET) with a gate length of 13 nm having a
> *c*-axis aligned crystalline In-Ga-Zn oxide (CAAC-IGZO) channel
> was fabricated. The CAAC-IGZO FET has an off-state leakage
> current of 200 yA/μm, a cutoff frequency of 60GHz, and a
> maximum oscillation frequency of 16GHz. A CAAC-IGZO FET,
> though it is a small transistor, withstands voltages up to
> approximately 2.5 V. It also has stable current characteristics with
> less temperature dependence than Si devices. We have constructed
> an equivalent-circuit model of the CAAC-IGZO FET and designed
> an RF amplifier to show CAAC-IGZO FET's applicability to the
> GHz-range RF circuitry.

INTRODUCTION

A field-effect transistor (FET) using a crystalline oxide semiconductor material that has alignment in the *c*-axis direction (CAAC-IGZO) is applied to various electronic circuits (1-8, 27). In the LSI field, it is applied to non-volatile and analog memories utilizing its extremely low off-state leakage current (9-11). CAAC-IGZO FETs can be embedded in the back-end-of-line (BEOL) layers of complementary MOS (CMOS) logic processes. This means that circuit blocks having supply voltages and/or functions different from those of CMOS circuitry can be constructed on top of the CMOS circuitry (Fig. 1). Furthermore, CAAC-IGZO FETs may be used as an access transistor of resistive random access memory (RRAM) (12) and magnetoresistive random access memory (MRAM).

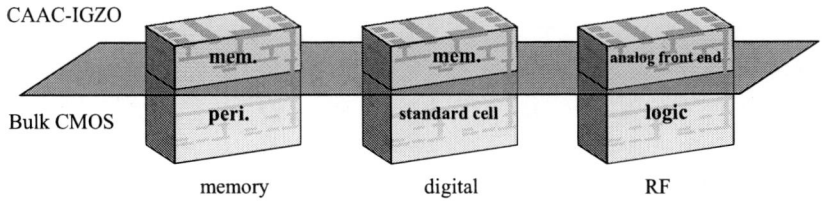

Fig. 1. Conceptual sketch of a CAAC-IGZO-on-CMOS stack.

However, exploration of CAAC-IGZO FET's application in radio-frequency (RF) space has so far been limited since CAAC-IGZO FETs' mobility is lower and their gate lengths longer than those of Si devices (13-15). Circuits that have been proposed so far still are in the megahertz range (16-19).

It can be said that CAAC-IGZO FETs still have room for scaling, as CAAC-IGZO has an energy bandgap that is wider than that of silicon, and a CAAC-IGZO FET has a source-to-drain voltage tolerance higher than that of Si devices. Scaling the CAAC-IGZO FETs would enable circuits faster than those already presented, opening wider application possibilities of the CAAC-IGZO FET. In addition, utilizing its stackability with CMOS circuitry, CAAC-IGZO FETs can provide a value-adding 3D circuitry that contributes to smaller RF circuits (and hence smaller packages) that are demanded for IoT chips (20, 21).

This work introduces the fabrication and characterization of scaled CAAC-IGZO FETs for application to high-speed circuitry and the construction of its equivalent-circuit model. Furthermore, an RF amplifier was designed using the equivalent-circuit model to study the feasibility of GHz-range RF circuits with CAAC-IGZO FETs, the results of which are also reported in this work.

FUNDAMENTAL CHARACTERISTICS OF CAAC-IGZO FET

A Hall-effect measurement of a CAAC-IGZO film was performed. When the CAAC-IGZO FET is assumed to be embedded into the BEOL layers of CMOS logic processes, the maximum process temperature of the CAAC-IGZO FET is preferably 400°C. Therefore, the Hall-effect mobility of CAAC-IGZO films having different carrier densities after a post-deposition anneal at 400°C was measured depending on their carrier densities. As shown in Fig.2, it can be seen that a CAAC-IGZO film would exhibit Hall-effect mobilities of 8.7 to 21.5 cm^2/Vs.

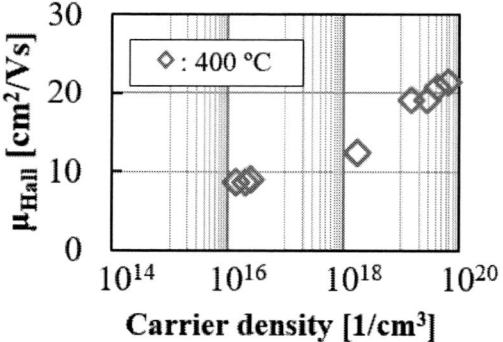

Fig. 2. Hall-effect measurement on a CAAC-IGZO film.

PROCESS FLOW AND STRUCTURE OF CAAC-IGZO FET

The process flow and a schematic view of the CAAC-IGZO FET are shown in Figs. 3 and 4, respectively. The cross-sectional Scanning transmission electron microscope (STEM) images of the scaled CAAC-IGZO FET fabricated for this work are shown in Fig. 5(a) for the channel length direction, and in Fig. 5(b) for the channel width direction.

The CAAC-IGZO FET of this work has a trench-gate self-aligned (TGSA) structure (22). To improve the current supply capability of the CAAC-IGZO FET, the device dimensions have been altered from a scaled device in the previous work (22). The CAAC-IGZO FET of this work has a gate length of 13.0 nm as shown in Fig. 5(a),(21.4 nm in the previous work), a gate width of 26 nm as in Fig. 5(b), and a gate dielectric thickness of 3.3 nm (6.0 nm in the previous work) in equivalent oxide thickness (EOT).

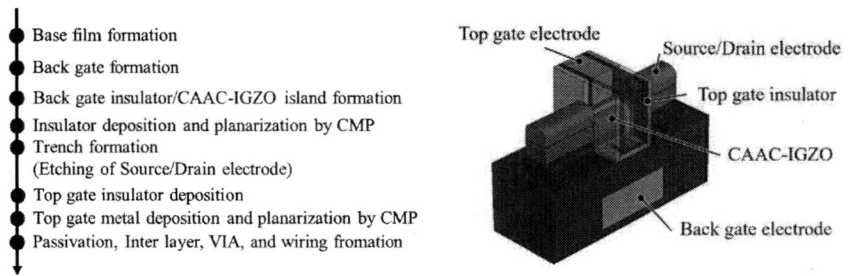

- Base film formation
- Back gate formation
- Back gate insulator/CAAC-IGZO island formation
- Insulator deposition and planarization by CMP
- Trench formation
  (Etching of Source/Drain electrode)
- Top gate insulator deposition
- Top gate metal deposition and planarization by CMP
- Passivation, Inter layer, VIA, and wiring fromation

Fig. 3. FET process flow.

Fig. 4. Schematic view.

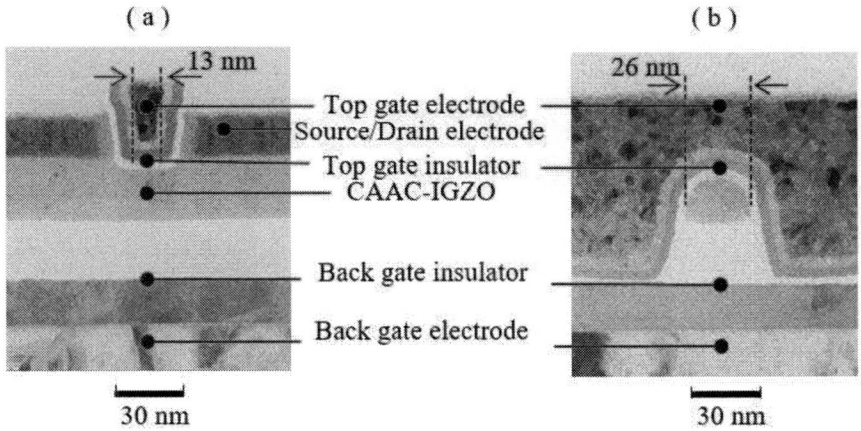

Fig. 5 Cross sections of CAAC-IGZO FET.
(a) channel length direction.　　(b) channel width direction.

STATIC CHARACTERISTICS OF CAAC-IGZO FET

Back gate bias dependence and temperature dependence of the I_d-V_{gs} characteristics and the transconductance g_m of a CAAC-IGZO FET having a dimension W/L = 26 nm/13 nm are shown in Figs. 6(a) and 6(b), respectively. In this measurement, the drain voltage V_{ds} is fixed at 0.9 V and the gate voltage V_{gs} is swept from 0 V to 2.5 V. The back gate bias dependence is measured under room temperature, with the back gate biases from −6 V to

6 V incremented in 2 V steps. The temperature dependence is measured with V_{bs} fixed at 0 V and under four temperature conditions (−40°C, 27°C, 85°C, and 150°C).

The back gate bias dependence measurement results show that moving the back gate bias of the CAAC-IGZO FET from +6 V to −6 V moves its threshold voltage from 0.94 V to 1.70 V. The temperature dependence measurement results show that the g_m that reflects the mobility of the device increases as the temperature increases (Fig. 6(b)). The temperature dependence of CAAC-IGZO FETs' I_d is the opposite of that of CMOS devices, as the field-effect mobility of CMOS devices decreases with rising temperatures.

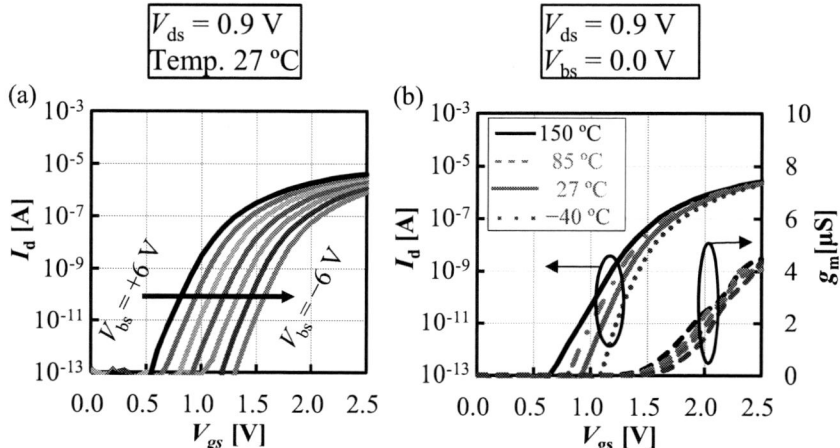

Fig. 6 (a) V_{bs} dependence and (b) temperature dependence of I_d-V_{gs} curves of CAAC-IGZO FET (W/L = 26 nm/13 nm).

Fig. 7 shows the gate voltage dependence of the I_d-V_{ds} characteristics of the same CAAC-IGZO FET. The drain voltage V_{ds} was swept from 0 V to 2.5 V with gate voltage Vgs conditions set from 1.1 V to 2.5 V in 0.2 V steps under room temperature. The drain current of the device did not clearly show saturation even with a drain voltage of up to 2.5 V.

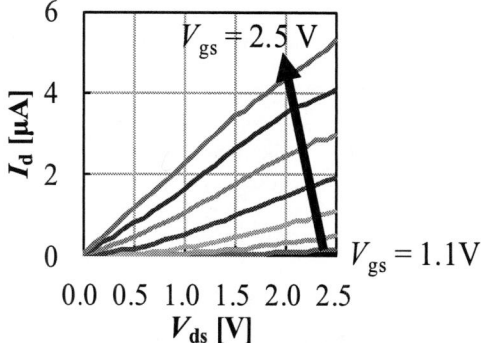

Fig. 7 I_d-V_{gs} curves of CAAC-IGZO FET (W/L = 26 nm/13 nm).

One of the defining characteristics of the CAAC-IGZO FET is its extremely low off-state leakage current shows the leakage current of the CAAC-IGZO FET measured with a custom circuit (22) Fig. 8. The size of the CAAC-IGZO FET were W/L = 520 μm/13 nm (M = 20,000 W/L = 26 nm/13 nm). D-S and D-TG mean drain-source leakage current and drain-top gate leakage current, respectively. Unit prefixes for the leakage current added to the y-axis of the plot are atto- (a), zepto- (z), and yocto- (y), which denote 10^{-18}, 10^{-21}, and 10^{-24}, respectively.

The estimated drain-source leakage current at room temperature is 200 yA/μm, higher than 5 yA/μm shown in a previous work (22). This is due to channel length and gate dielectric thickness scaling, but the leakage is still extremely low. In addition, the drain-top gate leakage current is smaller than that of the drain-source by approximately 2 digits, showing that the contribution of leakage through the top gate is minimal.

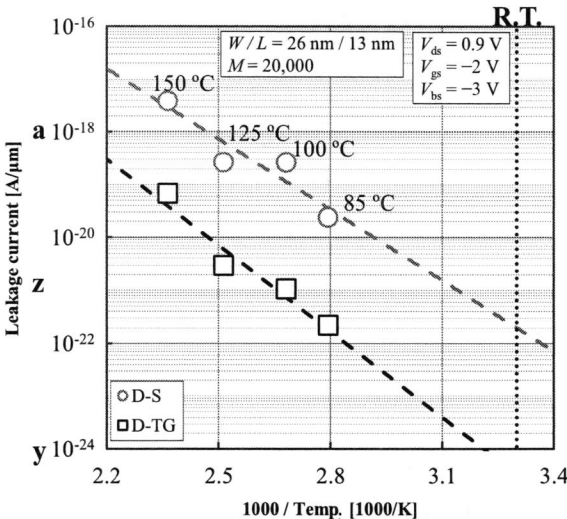

Fig. 8 Leakage current of a CAAC-IGZO FET.

Figs. 9 and 10 are contour plots showing the gate voltage-drain voltage dependence of transconductance g_m and drain conductance g_d.

It can be seen that CAAC-IGZO FET's g_m reaches a maximum value around the vicinities of a voltage condition at $V_{gs} = 2.5$ V and $V_{ds} = 2.5$ V, while the g_d shows a different tendency, reaching its maximum value around the vicinities of a voltage condition at $V_{gs} = 2.6$ V and $V_{ds} = 1.4$ V. To utilize a transistor as an amplifier, a voltage condition with high g_m efficiency needs to be used.

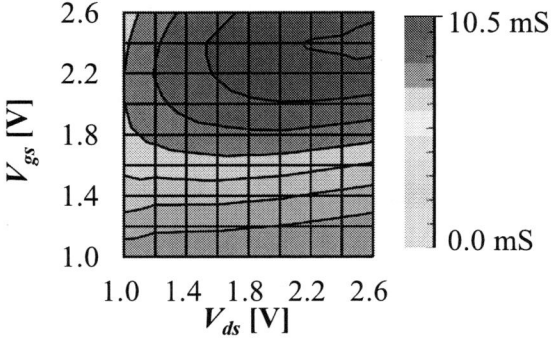

Fig. 9 V_{ds} and V_{gs} dependence of g_m of CAAC-IGZO FET with $W/L = 34.944$ μm/13 nm.

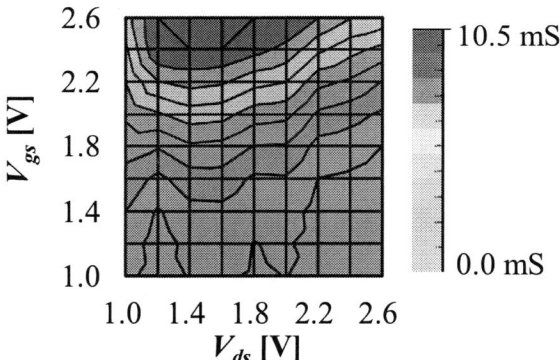

Fig. 10 V_{ds} and V_{gs} dependence of g_d of CAAC-IGZO FET with $W/L = 34.944$ μm/13 nm.

HIGH FREQUENCY PERFORMANCE OF CAAC-IGZO FET

High-frequency characteristics of the CAAC-IGZO FET were evaluated with a network analyzer. We used a Keysight Technologies N5247A network analyzer and a bias tee that provides a DC bias. For this measurement, three test element groups (TEGs) were measured. The first TEG (DUT-TEG) is one having the CAAC-IGZO FET devices of interest. The second TEG is an open TEG that has the CAAC-IGZO FETs removed from the first TEG. The third TEG is the first TEG with the ports short-circuited. The parasitic components in the measurement results of the first TEG are removed through open-short de-embedding to calculate the intrinsic high-frequency characteristics of the CAAC-IGZO FET.

Figs. 11 and 12 are contour plots showing the gate voltage-drain voltage dependence of two performance indices of a transistor, cutoff frequency f_T and maximum oscillation frequency f_{max}. The size of the CAAC-IGZO FET were W/L = 34.944 μm/13 nm (1344 W/L = 26 nm/13 nm devices connected in parallel). The voltages applied were V_{ds} of 1.0 V to 2.6 V, V_{gs} of 1.0 V to 2.6 V, and the temperature was set at 27°C.

Parameter f_T, which represents the current gain of the device. The maximum value of f_T was 63.7 GHz obtained when V_{gs} = 2.4 V and V_{ds} = 2.6 V.

Parameter f_{max}, which represents power gain of the device and attains a maximum value with a voltage condition that result in higher g_m (as shown in Figs. 12 and 13), i.e., with comparatively high V_{gs} and V_{ds}. The maximum value of f_{max} was 17.7 GHz obtained when V_{gs} = 2.4 V and V_{ds} = 2.4 V.

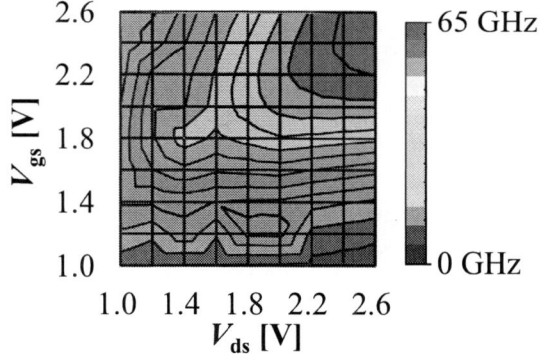

Fig. 11 V_{ds} and V_{gs} dependence of f_T of CAAC-IGZO FET with W/L = 34.944 μm/13 nm.

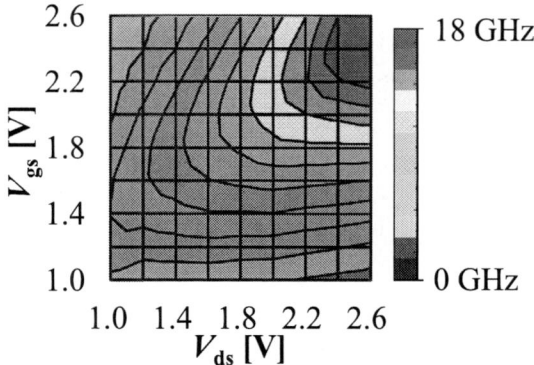

Fig. 12 V_{ds} and V_{gs} dependence of f_{max} of CAAC-IGZO FET with W/L = 34.944 μm/13 nm.

Figs. 13(a) and 13(b) show the current gain |H21| and the unilateral gain |Gu| plotted against the input frequency at a voltage condition ($V_{gs} = 2.5$ V and $V_{ds} = 2.5$ V) that enables high values in both f_T and f_{max}. A cutoff frequency $f_T = 60$ GHz and a maximum oscillation frequency $f_{max} = 16$ GHz were obtained. These results show that the CAAC-IGZO FET is a device that has ample capabilities in both current and voltage gains in frequencies ranging on the order of a few gigahertz.

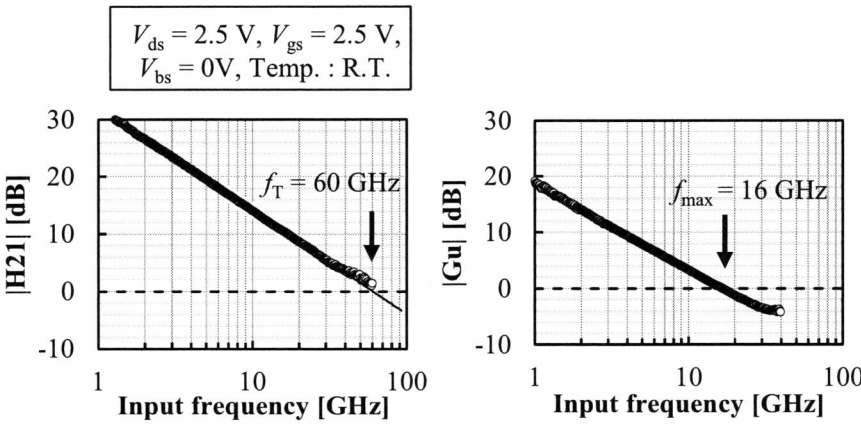

Fig. 13 CAAC-IGZO FET with W/L= 34.944 μm/13 nm.
  (a) Current gain at the optimum voltage (b) Unilateral gain at the optimum voltage

CAAC-IGZO FETs' field-effect mobility does not have a large temperature dependence (23). The temperature dependence of f_T was evaluated to check whether similar temperature characteristics would be obtained during high-frequency operation. Dimensions of the CAAC-IGZO FET were $W/L = 34.944$ μm/13 nm (1344 W/L = 26 nm/13 nm devices connected in parallel). The voltages applied were $V_{ds} = 2.5$ V, $V_{gs} = 2.5$ V, and the temperature ranged from −40°C to +150°C. As shown in Fig. 14, the f_T results normalized with the result at 25°C shows a positive correlation of f_T with temperature.
The f_T change in the temperatures ranging from −40°C to 150°C is 11.7%, which suggests that the temperature dependence of g_m (Fig. 7(b)) is significantly influencing the f_T results. A device that retains its performance at higher operating temperatures is highly welcomed by circuit designers.

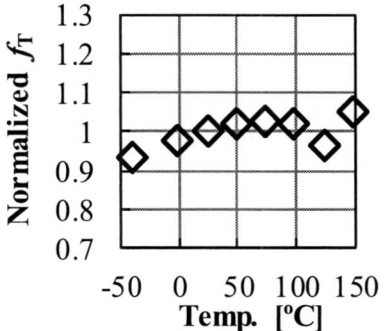

Fig. 14 Temperature dependence of f_T normalized at 25°C of CAAC-IGZO FET with W/L= 34.944 μm/13 nm

Trend seen in CAAC-IGZO FET's high-frequency characteristics

The trend in CAAC-IGZO FET's high-frequency characteristics is shown in Fig. 15. It is generally known that f_T of an FET increases when the FET's gate length is scaled down. The f_T of the CAAC-IGZO FET fabricated for this work is the highest figure at the time of this work.

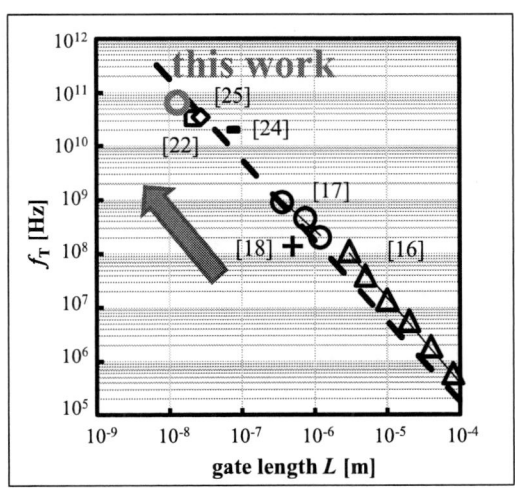

Fig. 15 Comparison of f_T and gate length of various CAAC-IGZO FETs.

Equivalent-circuit model

In order to design a circuit that operates at a frequency range of a few gigahertz. we created an equivalent circuit model that matches the high frequency characteristics of the CAAC-IGZO FET.As shown in Fig. 20, the equivalent-circuit model was composed of a) pad component, b) interconnect component, and c) intrinsic FET component, and a parameter fitting of to the measured data was performed.

The TEG for network measurement has a relatively large parasitic capacitance originating from the measurement pads and metal routing. Hence, an equivalent circuit model for the open TEG was also constructed and fit to the measurement data, to check the validity of the model also for the parasitic components. (Figs. 16 and 17)

A back gate terminal would also be needed for this model, however for this work we assumed the V_{bs} as fixed at 0 V and eliminated this parameter.

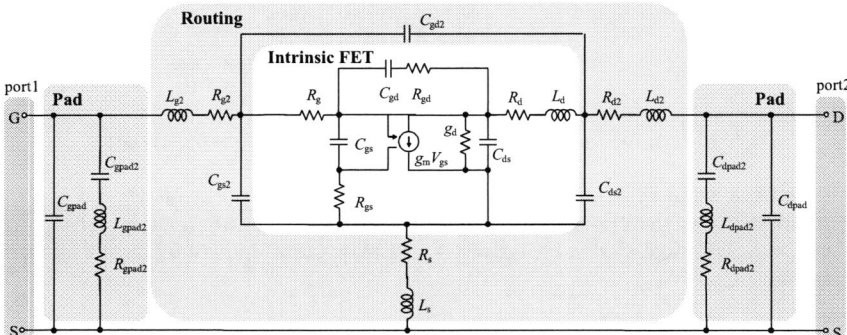

Fig. 16 DUT-TEG model

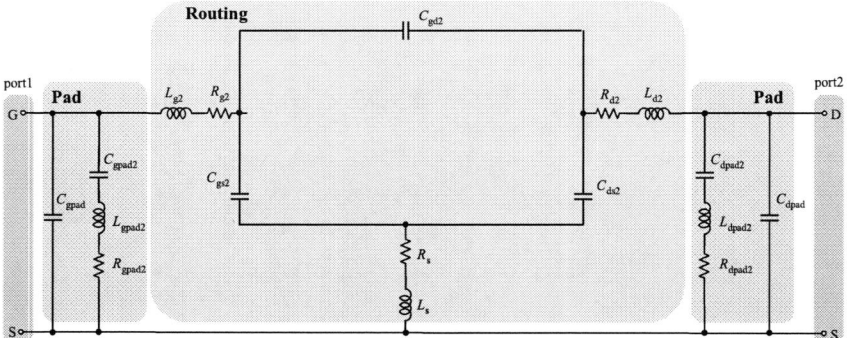

Fig. 17 OPEN-TEG model

Figs. 18 and 19 compare the S-parameters (Scattering parameters) obtained from equivalent-circuit models of DUT TEG and open TEG, and the S-parameters obtained from measurement data. Arrows in the figure show the direction in which the frequency becomes higher.

Though there is some deviation in high-frequency regions, the model fits the measured results well enough for designing circuits that operate at a frequency range of a few gigahertz. The reason for the deviation is believed to be originating from the accuracy of the de-embedding.

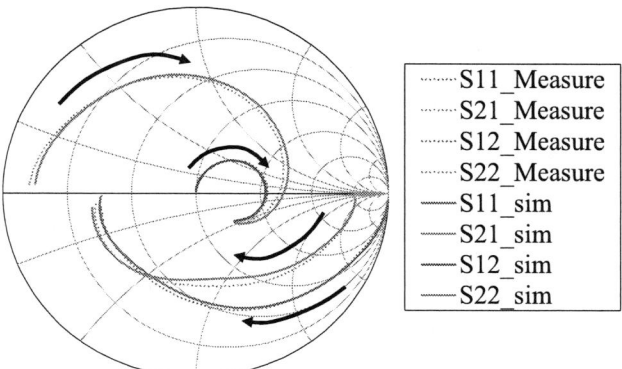

Fig. 18 DUT-TEG S-parameters 0.1GHz ~ 60.1GHz
Simulation model vs Measurement

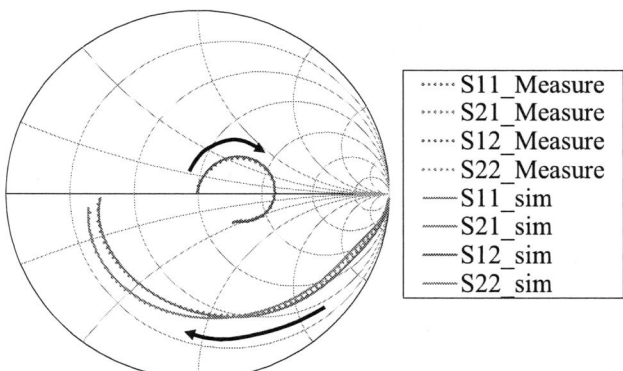

Fig. 19 OPEN-TEG S-parameters 0.1GHz ~ 60.1GHz
Simulation model vs Measurement

The equivalent circuit of the intrinsic FET component is shown in Fig. 20. This is a model that adds source resistance, gate resistance, and drain inductance to the hybrid-pi equivalent circuit model conventionally used for transistor modeling (26). The g_m and g_d values fit well with the measured static characteristics, showing that the extracted parameters are reasonable.

Intrinsic FET

Cgs	23.2	fF
Cgd	26.8	fF
Cds	5.5	fF
gm	10.6	mS
gd	2.14	mS
Rg	30	Ω
Rd	9	Ω
Rgs	185	Ω
Rgd	225	Ω
Ld	0.0534	nH

Fig. 20 Small-signal equivalent circuit of
CAAC-IGZO FET with W/L=34.944 µm/13 nm.

Application of CAAC-IGZO FET to RF amplifier

Using the small-signal equivalent circuit above, a common-source RF amplifier having matching circuits in its input and output was designed (Fig. 21). An RF amplifier that is capable to operating at frequencies of 2 to 3 gigahertz was simulated, adjusting the inductor- and capacitor-related parameters to obtain the required gain. Bias voltages V_{bias1} and V_{bias2} are both set at 2.5 V. Gain(S21) was modeled to be 11.5 dB at a frequency of 2.8GHz, showing that a GHz-range RF circuit can be composed of CAAC-IGZO FETs.

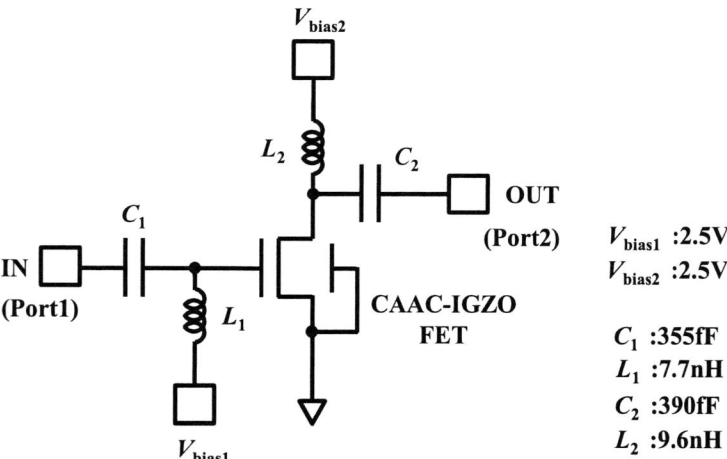

V_{bias1} :2.5V
V_{bias2} :2.5V

C_1 :355fF
L_1 :7.7nH
C_2 :390fF
L_2 :9.6nH

Fig. 21 One-Stage CAAC-IGZO FET Amplifier circuit

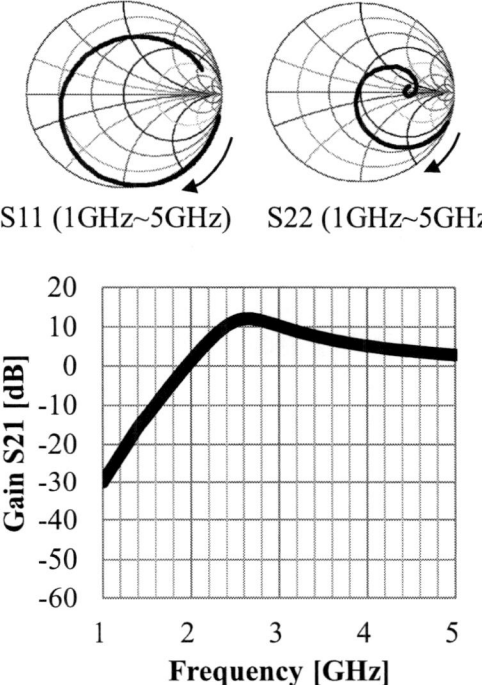

S11 (1GHz~5GHz)   S22 (1GHz~5GHz)

Fig. 22 One-Stage CAAC-IGZO FET Amplifier.
S-parameters (S11), (S22) Smith chart and Gain S21(dB)

CONCLUSION

An FET with a gate length of 13 nm having a *c*-axis aligned crystalline In-Ga-Zn oxide (CAAC-IGZO) channel was fabricated. Our CAAC-IGZO FET exhibited an off-state leakage current of 200 yA/μm, a cutoff frequency of 60GHz, and a maximum oscillation frequency of 16GHz. Its cutoff frequency changes by 11.7% under a temperature range of −40°C to 150°C, and its change has a positive correlation to temperature. This work showed that the CAAC-IGZO FET can be applied to circuits that operate at a few gigahertz, and that the application range of the CAAC-IGZO FET can be widened.

Acknowledgement

The authors gratefully acknowledge the contributions of Prof. Nakura of Fukuoka University to this paper.

References

1. Shunpei Yamazaki, Hideomi Suzawa, Koki Inoue, Kiyoshi Kato, Takuya Hirohashi, Kenichi and Okazaki, Noboru Kimizuka, *JJAP*, vol.53, pp. (04ED181)-(04ED1810), (2014).
2. Takahiko Ishizu, Yuto Yakubo, Kazuma Furutani, Atsuo Isobe, Masashi Fujita, Tomoaki Atsumi, Yoshinori Ando, Tsutomu Murakawa, Kiyoshi Kato, Masahiro Fujita, and Shunpei Yamazaki, *VLSI*, pp.C48-C49, (2019).
3. Munehiro Kozuma, Yuki Okamoto, Takashi Nakagawa, Takeshi Aoki, Yoshiyuki Kurokawa, Takayuki Ikeda,Yoshinori Ieda, Naoto Yamade, Hidekazu Miyairi, Makoto Ikeda, Masahiro Fujita, and Shunpei Yamazaki, *T-VLSI*, vol. 25, Issue 1, pp. 125-138, (2017).
4. Shuichi Katsui, Hidetomo Kobayashi, Takashi Nakagawa, Yuki Tamatsukuri, Hideaki Shishido, Shogo Uesaka, Ryohei Yamaoka, Takaaki Nagata, Tomoya Aoyama, Kosei Nei, Yutaka Okazaki, Takayuki Ikeda, and Shunpei Yamazaki, SID, pp. 311-314, (2019).
5. Tatsuya Onuki, Wataru Uesugi, Atsuo Isobe, Yoshinori Ando, Satoru Okamoto, Kiyoshi Kato, Tri Rung Yew,J. Y. Wu, Chi Chang Shuai, Shao Hui Wu, James Myers, Klaus Doppler, Masahiro Fujita, and Shunpei Yamazaki, *JSSC*, Vol. 52, No. 4, pp. 925-932, (2017).
6. Takanori Matsuzaki, Tatsuya Onuki, Shuhei Nagatsuka, Hiroki Inoue, Takahiko Ishizu, Yoshinori Ieda, Naoto Yamade, Hidekazu Miyairi, Masayuki Sakakura, Tomoaki Atsumi, Yutaka Shionoiri, Kiyoshi Kato, Takashi Okuda, Yoshitaka Yamamoto, Masahiro Fujita, Jun Koyama, and Shunpei Yamazaki, *ISSCC*, pp.306-307, (2015).
7. Shuhei Maeda, Satoru Ohshita, Kazuma Furutani, Yuto Yakubo, Takahiko Ishizu, Tomoaki Atsumi, Yoshinori Ando, Daisuke Matsubayashi, Kiyoshi Kato, Takashi Okuda, Masahiro Fujita, and Shunpei Yamazaki, , *ISSCC*, pp.306-307, (2015).
8. Abidur Rahaman, Yuanfeng Chen, MD. Mehedi Hasan, and Jin Jang, *J-EDS*, vol. 7, pp.655-661, (2019).
9. Hiroki Inoue, Takeshi Aoki, Fumika Akasawa, Toshiki Hamada, Toshihiko Takeuchi, Kousei Nei, Takako Seki, Yuto Yakubo, Kei Takahashi, Shuji Fukai, Takahiko Ishizu, Munehiro Kozuma, Ryota Tajima, Takanori Matsuzaki, Takayuki Ikeda, Makoto Ikeda, and Shunpei Yamazaki, , *ISSCC*, pp.204-205, (2019).
10. Toshiki Hamada, Toshihiko Takeuchi, Takeshi Aoki, Munehiro Kozuma, Takayuki Ikeda, Makoto Ikeda, and Shunpei Yamazaki, SID, pp.1436-1439, (2019).
11. Yunpeng Li, Jin Yang, Yiming Wang, Pengfei Ma, Yvzhuo Yuan, Jiawei Zhang, Zhaojun Lin, Li Zhou, Qian Xin, and Aimin Song, *EDL*, vol.39, No. 2, pp.208-211, (2019).
12. Jixuan Wu, Fei Mo, Takuya Saraya, Toshiro Hiramoto, and Masaharu Kobayashi, *Symp.* VLSI Tech. Dig., THL.4 (2020)
13. C. Tanaka and Keiji Ikedaet, *ICMTS*, pp.23-26, (2018).
14. Matteo Ghittorelli, Fabrizio Torricelli, Carmine Garripoliy, Jan-Laurens J.P. van der Steenz, Gerwin H. Gelinckzx, Sahel Abdiniay, Eugenio Cantatorey, and Zsolt M. Kov´acs-Vajna, *ESSDERC*, pp.98-101, (2017).

15. Wanling Deng, Jielin Fang, Xixiong Wei, Weijing Wu, and Junkai Huang, *T-ED*, vol. 65, No. 4, pp.1370-1376, (2018).
16. Pydi Ganga Bahubalindruni, Asal Kiazadeh, Allegra Sacchetti, Jorge Martins, Ana Rovisco, Vítor Grade Tavares, Rodrigo Martins, Elvira Fortunato, and Pedro Barquinha, EDL vol.12, No. 6, pp.515-518, (2016).
17. Yiming Wang, Jin Yang, Hanbin Wang, Jiawei Zhang, He Li, Gengchang Zhu, Yanpeng Shi, Yuxiang Li, Qingpu Wang, Qian Xin, Zhongchao Fan, Fuhua Yang, and Aimin Song, *T-ED*, pp.1377-1382, (2018).
18. K. Ishida, T. Meister, R. Shabanpour, B. K. Boroujeni, C. Carta, G. Cantarella, L. Petti, N. Münzenrieder, G. A. Salvatore, G. Tröster, and F. Ellinger, AM-FPD, pp.273-276, (2016).
19. Nikolas Papadopoulos, Steve Smout,Myriam Willegems, Marc Ameys, Ganesh Rathinavel, Gerd Beeckman, Jan Stuijt, and Kris Myny, *IFETC*, pp.1-2, (2018).
20. Chien-Fu Tseng, Chung-Shi Liu, Chi-Hsi Wu, and Douglas Yu, *J-EDS*, vol. 7, pp.495-502, (2019).
21. Maryam Tabesh, Nemat Dolatsha, Amin Arbabian, Ali M. Niknejad, *ECTC*, pp.1-6, (2016).
22. Hitoshi Kunitake, Kazuaki Ohshima, Kazuki Tsuda, Noriko Matsumoto, Tatsuki Koshida, Satoru Ohsita, Hiromi Sawai, Yuichi Yanagisawa, Shiori Saga, Ryo Arasawa, Takako Seki, Ryunosuke Honda, Haruyuki Baba, Daigo Shimada, Hajime Kimura, Ryo Tokumaru, Tomoaki Atsumi, Kiyoshi Kato, and Shunpei Yamazaki, *J-EDS*, vol. 7, pp.495-502, (2019).
23. Chien-Fu Tseng, Chung-Shi Liu, Chi-Hsi Wu, and Douglas Yu, *ECTC*, pp.1-6, (2016).
24. Yoshinobu Asami, Akihisa Shimomura, Yutaka Okazaki, Daisuke Matsubayashi, Masashi Tsubuku, Motomu Kurata, Satoru Okamoto, Shinya Sasagwa, Tomoaki Moriwaka, Tetsuya Kakehata, Yuto Yakubo, Kiyoshi Kato, Yoshitaka Yamamoto, and Shunpei Yamazaki ,*SSDM*, pp.1084-1085, (2015).
25. D. Matsubayashi, Y. Asami, Y. Okazaki, M. Kurata, S. Sasagawa, S. Okamoto, Y. Iikubo, T. Sato, Y. Yakubo, R. Honda, M. Tsubuku, M. Fujita, T. Takeuchi, Y. Yamamoto, and S. Yamazaki *IEDM*, pp.141-144, (2015).
26. L. J. Giacoletto, *JSSC*, vol. 4, issue 2, pp.80-83, (1969).
27. Shunpei Yamazaki, Satoru Ohshita, Masashi Oota, Haruyuki Baba, Tatsuya Onuki, Hitoshi Kunitake, Kazuaki Ohshima, Daigo Shimada, Hajime Kimura, Tsutomu Murakawa, Tomoaki Atsumi, and Kiyoshi Kato, *Ceramic Engineering & Science* Volume1, Issue1 pp 6-20 , (2019).

Highly Reliable Metal Oxide Thin Film Transistors for Flexible Devices

Yukiharu Uraoka, Dianne C. Corsino, Juan Paolo S. Bermundo,
Mami N. Fujii, Mutsunori Uenuma

Division of Materials Science, Graduate School of Science and Technology, Nara
Institute of Science and Technology, 8916-5 Takayama, Ikoma, Nara 630-0192, Japan

Fully solution-processed amorphous InZnO (*a*-IZO) thin-film transistors (TFTs) were demonstrated utilizing solution-based channel, gate insulator, and electrodes with a maximum fabrication temperature of 300 °C. Particularly, a single layer of *a*-IZO was used as both the channel and the source/drain electrode layer by selectively tuning the role of *a*-IZO as a semiconductor or a conductor through photo-assisted treatments. By employing a self-aligned TFT structure, the *a*-IZO electrodes were functionalized by UV irradiation and excimer laser annealing (ELA). The fully solution-processed *a*-IZO TFTs exhibited high performance with an average mobility of up to 38 cm^2 V^{-1} s^{-1}, which surpasses those of previously reported approaches for fully solution-processed oxide TFTs. Moreover, the overall device performance, including subthreshold swing of 225 mV dec^{-1} and on-voltage of -0.4 V, is comparable to those of vacuum-processed oxide TFTs.

INTRODUCTION

For more than a decade, researches on thin-film transistor (TFT) technology have intensively focused on amorphous oxide semiconductors (AOSs) such as amorphous InGaZnO (*a*-IGZO) and amorphous InZnO (*a*-IZO). AOSs offer numerous merits such as high mobility, low-temperature processability, solution process compatibility, and wide band gap which improves transparency.[1] AOS-based TFTs have been well established to result in unquestionably high performance and highly reliable devices when fabricated using vacuum techniques that they are now used to drive various display panels such as active matrix liquid crystal displays and OLED TVs.[2] However, the rapidly advancing field of active-matrix electronics demands both cheaper manufacturing costs and high-throughput production. This cannot be achieved with the present dependence on vacuum technologies which are costly and have high energy requirement. To fully realize cost-efficient and large-area production, all layers need to be fabricated through solution process. One of the challenges in fabricating a fully solution-processed device is the incorporation of a solution-based conductive film for the electrodes. There are notable advances demonstrating the use of solution-based conductive films such as carbon-based materials (carbon nanotubes and graphene oxide),[3] conducting polymers and organic materials,[4] and metal nanostructure meshes.[5] However, these materials possess unsatisfactory conductivity, insufficient operational stability, and poor contact with oxide semiconductors which disqualify the possibility of its successful integration to the current electronic systems.[6] On the other hand, transparent conductive oxides, the most popular of which is ITO, are starting to get recognized due to its superior contact with AOSs and satisfactory conductivity for device applications.[7]

Here, we report a low-temperature approach for fully solution-processed oxide TFT fabrication using a-IZO for the channel and electrode materials, and fluorinated-polysilsesquioxane (F-PSQ) for the gate insulator. The maximum process temperature employed during film deposition steps was 300 °C and the required temperature during photo-assisted treatments was only 115 °C.

In this work, we elucidate the mechanism of role tuning of solution-processed a-IZO through a more complete and comprehensive film and device analyses. The performance and stability of the fully solution-processed a-IZO TFTs is also maximized by exploring the effect of fluorinated gate insulator and investigating various parameters of photo-assisted techniques, such as UV irradiation time, laser fluence (F), and ELA treatment environment. In particular, a treatment combination of UV irradiation and ELA in vacuum resulted in a-IZO TFTs with superior electrical characteristics such as excellent average μ = 38.0 cm^2 V^{-1} s^{-1}, subthreshold swing SS = 225 mV dec^{-1}, and on-voltage V_{on} (gate voltage (V_{gs}) at drain current (I_{ds}) = 1 nA) = -0.4 V. These results do not only show μ values that surpass the existing fully solution-processed TFTs but also exhibit device properties that are comparable with vacuum-processed oxide TFTs. The selective semiconductor-to-conductor transformation of a-IZO where localized modification of electrical properties of the oxide was also demonstrated which allows the use of a single a-IZO channel layer to act as both the electrode and channel layer. The selective conversion mechanism of a-IZO from a semiconductor into a conductor was elucidated by probing the effect of photo-assisted treatments on microstructural properties, chemical bonding, and composition of the a-IZO layer. Our findings suggest that a combination of UV and ELA results in an increased carrier concentration in the a-IZO through V_o generation and enhanced charge transport through densification and crystallization of a-IZO caused by laser-induced heating. This work reports another approach to develop transparent conductive oxide materials which will be beneficial in advancing technologies requiring promising optoelectronic properties.

RESULTS AND DISCUSSION

In devising an approach to fully solution-based TFT fabrication, material selection, device architecture, and appropriate treatment methods are vital considerations. Unlike the conventional TFT structure which requires various materials, numerous deposition and patterning steps, and demanding treatment methods, a simple oxide TFT fabrication employing a maximum process temperature of only 300 °C was developed as depicted in **Figure 1** and in the inset of **Figure 2**a. The self-aligned TFT structure permits the implementation of only three deposition steps and two patterning steps. We previously showed the elemental mapping of the cross-section of the fully solution-processed a-IZO TFT obtained by energy dispersive X-ray spectroscopy (EDX).[8] The results of EDX confirms the distinct IZO gate, F-PSQ GI, and IZO channel and source/drain layers. The In and Zn diffusion into the F-PSQ GI is observed to be minimal which verifies that the top and bottom a-IZO layers are completely isolated. Thus, it can be established that only the exposed a-IZO areas can be transformed into conductive electrodes by the photo-assisted treatments.

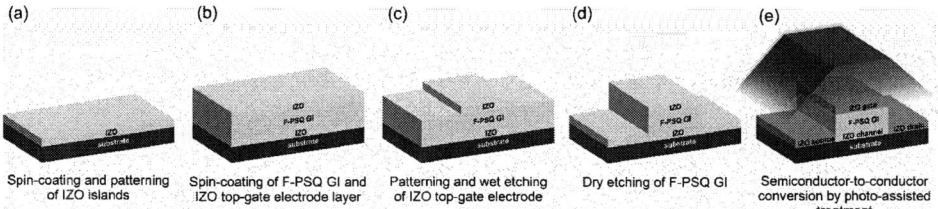

Figure 1. Schematic diagram of the simple fully solution-processed *a*-IZO TFT fabrication: (a) *a*-IZO deposition by spin-coating and patterning of the *a*-IZO islands, (b) deposition of F-PSQ GI and *a*-IZO top-gate electrode layer by spin-coating, (c) patterning the self-aligned structure and wet etching of *a*-IZO top-gate electrode, (d) dry etching of F-PSQ GI, and (e) functionalization of the *a*-IZO electrodes by photo-assisted methods.

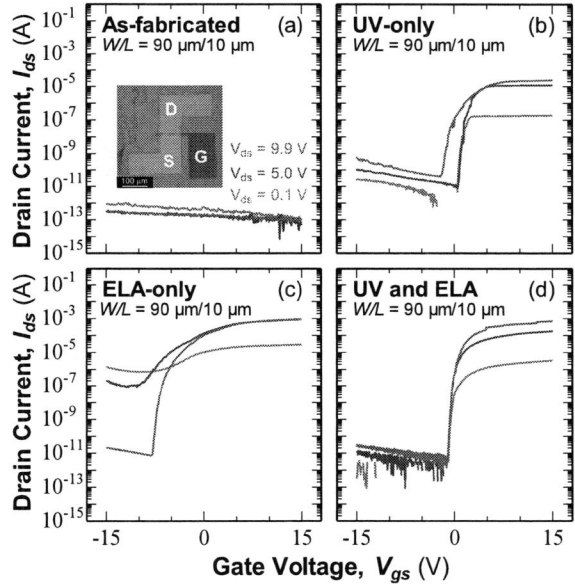

Figure 2. Transfer characteristics of the fully solution-processed *a*-IZO TFTs before (a) and after irradiation with (b) only UV, (c) only ELA, and (d) a combination of UV and ELA. The transfer characteristics were measured at V_{ds} = 0.1, 5.0, and 9.9 V. The *a*-IZO channel width (W) and length (L) is 90 μm and 10 μm, respectively. The inset in (a) shows the top-view of the fully solution-processed *a*-IZO TFTs.

Figure 2 shows the transfer characteristics of the fully solution-processed *a*-IZO TFTs before and after performing the photo-assisted treatments. Figure 2a illustrates the absence of TFT switching behavior in the as-fabricated *a*-IZO TFT due to the insufficient conductivity of the as-deposited *a*-IZO electrodes which still retain its semiconductor property. After performing the photo-assisted treatments, the *a*-IZO areas directly exposed to UV irradiation and ELA were functionalized into conductive electrodes. This results in the apparent switching in the transfer characteristics as shown in Figure 2b-d. Particularly, it can be observed that the combination of UV and ELA yielded smooth transfer curves with negligible V_{on} shifting and minimal degradation. The μ was then extracted from the transfer characteristics in the linear regime using the smallest V_{ds} of 0.1 V using Equation 1. Here, g_m is the transconductance measured by the semiconductor parameter analyzer, L and W is the channel length and width, respectively, C_{ox} is the capacitance of the gate insulator, and V_{ds} is the drain voltage.

$$\mu = g_m \frac{L}{WC_{ox}V_{ds}} \tag{1}$$

With regards to device stability, **Figure 3** illustrates the temporal degradation and cyclic measurement behavior of the fully solution-processed a-IZO TFTs irradiated with UV only and the combination of UV and ELA. As in Figure 3c, it can be observed that a single UV treatment results in the severe degradation in the I_{on} and SS after performing a cyclic test where the transfer characteristics were measured for ten consecutive times. In addition, the TFT switching behavior disappeared after only two weeks of storage at room temperature in dark air condition as illustrated in Figure 3b. In contrast, it is evident from Figure 3f that the a-IZO TFTs subjected to the combination of UV and ELA resulted in a more stable TFT performance showing minimal I_{on} degradation after the cyclic measurement. Moreover, even after three months of storage in the same conditions, the a-IZO TFTs irradiated with both UV and ELA still displayed ideal switching behavior but with decreased μ from 39.2 to 18.2 cm^2 V^{-1} s^{-1}.

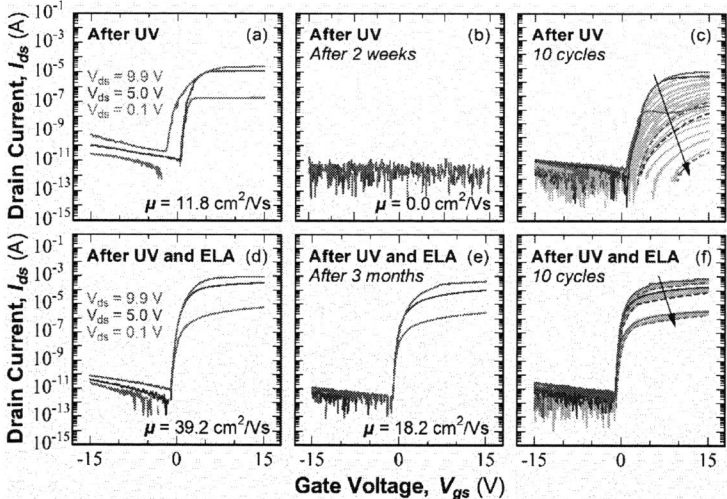

Figure 3. Stability of the fully solution-processed a-IZO TFTs. The transfer characteristics of the TFTs irradiated (a) with UV-only and (d) with a combination of UV and ELA are measured (b) after two weeks and (e) after three months of storage in air, respectively. (c,f)

The evolution of transfer characteristics of the a-IZO TFTs against cyclic measurement was obtained from ten consecutive cycles of forward sweep measurement.

To explicate the laser-induced heating effect on the self-aligned structure of the fully solution-processed a-IZO TFT employed in this work, simulation of temperatures was carried out. The results of the two-dimensional simulation (COMSOL Multiphysics) of temperatures during KrF excimer laser irradiation on the irradiated and non-irradiated a-IZO areas as well as on various interfaces in the self-aligned TFT structure in **Figure 4**. Figure 4a shows the simulated temperatures during ELA from 1.45 to 2.45×10^{-7} s which confirms that the laser-induced heating penetrates throughout the thickness of the exposed a-IZO electrode layers. Nevertheless, it can also be observed that the extent of heating, such as maximum temperature generated, as illustrated in Figure 4b-c, and heating duration, varies at different locations in the TFT structure. The exposed IZO top gate electrode reached the highest maximum temperature of 1780 K (1507 °C) while the exposed IZO source and drain electrodes reached 1329 K (1056 °C). On the other hand, the unexposed a-IZO channel area, which is shielded by the thick F-PSQ gate insulator and IZO top gate electrode layers, only reached a maximum temperature of 726 K (453 °C) which is substantially low for crystallization to occur. The extremely high temperature on the exposed IZO areas is likely the cause of the crystallization in the electrodes despite the rapid process. Meanwhile, the substantially lesser heating effect on the unexposed a-IZO channel area permits the retention of the amorphous nature of the a-IZO channel. The results of the simulation confirm the preservation of semiconducting properties and achievement of conducting properties in selected areas on the same a-IZO layer.

Interestingly, the duration of laser-induced heating is revealed to vary at different locations in the self-aligned TFT structure, particularly between the IZO gate area and the IZO source/drain areas as shown in Figure 4d-e. It was evaluated that the laser-induced heating in the irradiated IZO source/drain areas was maintained for $>1.0 \times 10^{-7}$ s until the temperature drops to $\approx$ 400 K. On the other hand, the IZO top gate electrode was heated for a considerably prolonged time. It is worth pointing out that the maximum temperature at the interface between SiO_2 and Si substrate is only 356 K (83 °C). Hence, ELA is demonstrated as a promising low-temperature treatment method for high throughput fully solution-processed TFT fabrication especially on flexible heat-sensitive substrates.

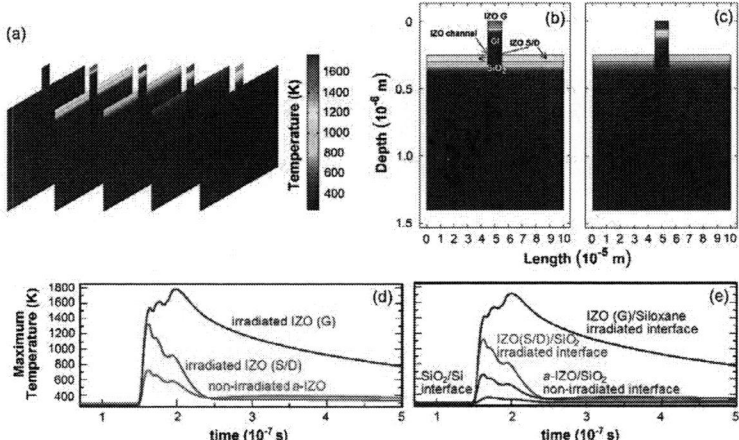

Figure 4. Two-dimensional simulation of laser-induced heat generation in the fully solution-processed a-IZO TFT (a) during ELA from 1.45 to 2.45 $\times$ 10^{-7} s, and at (b) 1.61 $\times$ 10^{-7} s and (c) 1.98 $\times$ 10^{-7} s, when the temperature at the IZO source/drain electrodes and IZO gate electrode, respectively, are at a maximum. (d) Comparison of the maximum simulated laser-induced temperature at different a-IZO areas and (e) at different interfaces in the fully solution-processed a-IZO TFT structure.

The microstructural changes in the irradiated and non-irradiated a-IZO locations were also analyzed through scanning transmission electron microscopy (STEM). Focused ion beam was used to prepare a very thin cross-section of the fully solution-processed IZO TFT. STEM observation was performed specifically on the IZO gate and source/drain electrodes, and the IZO channel as shown in **Figure 5**. The a-IZO channel was verified to retain its amorphous structure or short-range order which supports the assumption that the shielded IZO channel area is least affected by the laser-induced heating during ELA. On the other hand, Moiré patterns can be clearly observed extending throughout the thickness of the IZO gate and source/drain electrodes, confirming crystallization caused by intense heat generation during ELA. Particularly, the crystal plane distances appearing in the IZO gate were measured to be 2.90 Å and 5.09 Å, which can be attributed to In$_2$O$_3$ crystal planes (222) and (002), respectively. Moreover, it is interesting to mention that while crystal grains are obvious in the IZO source/drain electrodes, the crystal size is noticeably smaller than those observed in the IZO gate electrode.

This can be attributed to the extended heating in the IZO gate electrode than in the source/drain electrodes as shown by the results of the COMSOL simulation. It should be noted that the layer underneath the IZO gate and source/drain electrodes is F-PSQ and SiO$_2$, respectively. The lower thermal conductivity of F-PSQ (0.1–0.2 W m^{-1} K^{-1}) [9] than that of SiO$_2$ (1.1–1.4 W m^{-1} K^{-1}) [10] likely promoted the extended heating which, in turn, resulted in the larger crystals in the IZO gate electrode. Additionally, it was observed from STEM images that the thickness of the IZO source and drain areas is smaller than that of the a-IZO channel despite being in the same layer. This result suggests that IZO densification was achieved after performing the photo-assisted methods.

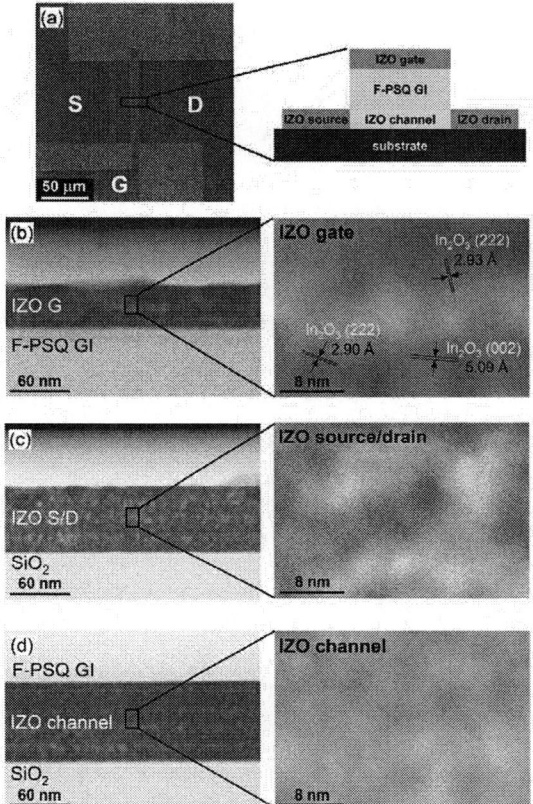

Figure 5. Scanning transmission microscopy (STEM) data of the different IZO areas in the fully solution-processed *a*-IZO TFT. (a) Top-view of the actual *a*-IZO TFT indicating the location where the STEM sample was extracted using FIB. The schematic of the cross-section is also shown. STEM images taken of the (b) IZO gate electrode, (c) IZO source/drain electrode, and (d) IZO channel layers and the corresponding high-magnification images of the bulk region.

Figure 6 illustrates the schematic diagram for the photo-assisted role tuning of *a*-IZO through semiconductor-to-conductor transformation. Initially, the as-deposited *a*-IZO has an amorphous nature and behaves as a semiconductor. TFT switching behavior cannot be observed in the transfer characteristics due to the insufficient conductivity of the as-deposited *a*-IZO. Meanwhile, enhancement in the conductivity of the electrodes were achieved after performing the photo-assisted treatments which can be explained by the changes in the chemical bonding and microstructure of *a*-IZO. After UV irradiation, IZO retained its amorphous nature while oxygen vacancies are created by breaking the In – O and Zn – O bonds. As a result, conductivity was enhanced, effectively transforming the

IZO into a conductor. After the additional ELA, IZO crystallization was induced due to laser heating which further enhances the charge transport in the IZO.

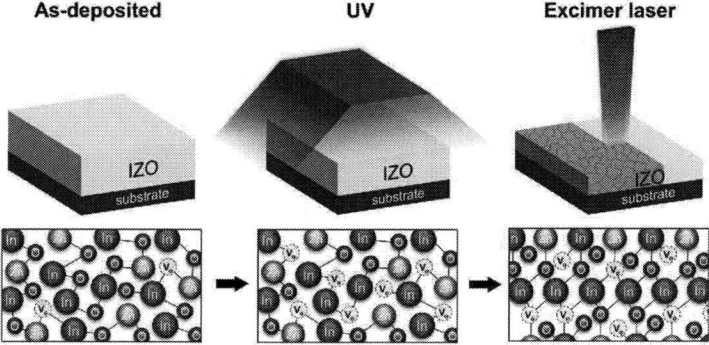

Figure 6. Schematic diagram of the low-temperature photo-assisted role tuning of a-IZO via semiconductor-to-conductor conversion using UV irradiation and KrF excimer laser annealing inducing oxygen vacancy generation and crystallization, respectively.

CONCLUSION

We have demonstrated a fully solution approach to oxide TFT fabrication by employing photo-assisted treatments namely UV irradiation and ELA. High performance fully solution-processed a-IZO TFTs with μ up to ≈ 40 cm^2 V^{-1} s^{-1}, V_{on} = -0.4 V, and SS = 225 mV dec^{-1} were achieved by a combination of UV irradiation for 60 min and ELA in vacuum at F = 120 mJ cm^{-2}, which are competitive with vacuum-processed oxide TFTs. It was also found that the combination of UV and ELA is necessary to achieve stable devices in terms of storage time and cyclic measurement. A single layer of a-IZO can be selectively tuned to act as both the channel and the source/drain electrodes through the semiconductor-to-conductor conversion of a-IZO in a self-aligned TFT structure. This functionalization process is expected to be applicable to many other materials for various device applications such as in 3D printing. Furthermore, the fully solution approach presented here can be applied to large-scale fabrication techniques for high-throughput roll-to-toll manufacturing.

ACKNOWLEDGMENT

The authors would like to thank Prof. H. Ikenoue in Kyushu University for his fruitful support. The authors also would like to thank Nissan Chemical Corporation for providing the InZnO precursor solutions and Kazuhiro Miyake for assisting with the sample preparation by FIB and STEM observation.

REFERENCES

1. Fortunato, E.; Barquinha, P.; Martins, R. Oxide Semiconductor Thin-Film Transistors: A Review of Recent Advances. *Adv. Mater.* **2012**, *24,* 2945–2986.
2. Xu, W.; Li, H.; Xu, J.-B.; Wang, L. Recent Advances of Solution-Processed Metal Oxide Thin-Film Transistors. *ACS Appl. Mater. Interfaces* **2018**, *10,* 25878–25901.
3. Feng, C.; Liu, K.; Wu, J.-S.; Liu, L.; Cheng, J.-S.; Zhang, Y.; Sun, Y.; Li, Q.; Fan, S.; Jiang, K. Flexible, Stretchable, Transparent Conducting Films Made from Superaligned Carbon Nanotubes. *Adv. Funct. Mater.* **2010**, *20,* 885–891.
4. Xia, Y.; Sun, K.; Ouyang, J. Solution Processed Metallic Conducting Polymer Films as Transparent Electrode of Optoelectronic Devices. *Adv. Mater.* **2012**, *24,* 2436–2440.
5. Kim, W.-K. Lee, S.; Lee, D. H.; Park, I. H.; Bae, J. S.; Lee, T. W.; Kim, J.-Y.; Park, J. H.; Cho, Y. C.; Cho, C. R.; Jeong, S.-Y. Cu Mesh for Flexible Transparent Conductive Electrodes. *Sci. Rep.* **2015**, *5,* 10715.
6. Bermundo, J. P. S.; Ishikawa, Y.; Fujii, M. N.; Ikenoue, H.; Uraoka, Y. Instantaneous Semiconductor-to-Conductor Transformation of a Transparent Oxide Semiconductor a-InGaZnO at 45 °C. *ACS Appl. Mater. Interfaces* **2018**, *10,* 24590–24597.
7. Ban, S.-G.; Kim, K.-T.; Choi, B. D.; Jo, J.-W.; Kim, Y.-H.; Facchetti, A.; Kim, M.-G.; Park, S. K. Low-Temperature Postfunctionalization of Highly Conductive Oxide Thin-Films toward Solution-Based Large-Scale Electronics. *ACS Appl. Mater. Interfaces* **2017**, *9,* 26191–26200.
8. Bermundo, J. P. S.; Kulchaisit, C.; Corsino, D. C.; Syairah, A.; Fujii, M. N.; Ikenoue, H.; Ishikawa, Y.; Uraoka, Y. High Performance All Solution Processed Oxide Thin-Film Transistor via Photo-induced Semiconductor-to-Conductor Transformation of a-InZnO. *SID Symposium Digest of Technical Papers* **2019**, *50,* 422–425.
9. Sugawara, A.; Takahashi, I. Thermal Conductivity of Polysiloxane in Intermediate Temperature Range. In *Thermal Conductivity 14;* Klemens P. G., Chu T. K., Eds. Springer: Boston, MA, 1976; pp 299–302.
10. Burzo, M. G.; Komarov, P. L.; Raad, P. R. Thermal transport properties of gold-covered thin-film silicon dioxide. *IEEE Trans. Compon. Packag. Technol.* **2003**, *26,* 80–88.

UV and Gate Stress Induced Defects in Amorphous Indium Gallium Zinc Oxide Thin Film Transistors and Self-Repair

Jingxin Jiang[a] and Yue Kuo*

Thin Film Nano & Microelectronics Research Laboratory, Texas A&M University, College Station, Texas 77843-3122, USA *yuekuo@tamu.edu

[a] Also with Information Science and Engineering, Shenyang University of Technology, Shenyang, China

> Defects were generated in the amorphous indium gallium zinc oxide thin film transistor under combined ultraviolet light illumination and gate bias stress conditions, which showed as the positive shift of the threshold voltage and the increase of the threshold swing. When exposed to the ultraviolent light without or with the positive gate voltage bias, positively charged defects were quickly generated and saturated at the gate dielectric interface. Majority of these defects were self-repaired at room temperature under air exposure within 5 minutes. On the other hand, under the ultraviolent exposure and negative gate voltage bias condition, two types of defects were generated, i.e., those loosely trapped at the interface and those deeply trapped in the bulk of the gate dielectric layer. At room temperature, the former were quickly self-repaired and the latter were slowly self-repaired. All defects were removed when annealed at 200°C. The understanding of the defect generation and removal mechanism is important for practical applications of this type of oxide thin film transistor.

Introduction

Amorphous indium gallium zinc oxide thin-film transistors (a-IGZO TFTs) are promising pixel driving devices for high performance flat-panel displays. Due to the delocalized s orbitals of heavy metal cations, which forms a largely dispersed conduction band with a small electron effective mass, the a-IGZO TFT has a high field effect mobility (μ_{eff}) (1). It is transparent to the visible light and can be fabricated at a low temperature (2-5). However, the a-IGZO TFT's instability induced by the light illumination and the gate voltage (V_{GS}) stress is a crucial concern in circuit and display applications (6-9). There are reports that the wavelength of the illumination light affects the a-IGZO TFT's subthreshold swing (S), threshold voltage (V_{th}), etc. (10,11).

Moreover, the light effect is enhanced with the increase of temperature due to the lowering of the barrier height at the source region (11). Most studies of the light illumination effect on the a-IGZO TFT are in the visible wavelength range. Recently, there are reports that the a-IGZO TFT can be applied to the ultraviolet (UV) light sensing or imaging (12-14). Since the UV exposure can cause the instability of the a-IGZO TFT (15-17), it is necessary to understand the defect generation and removal mechanism. In actual display or sensor applications, TFTs are often operated under the simultaneous light exposure and gate bias condition. The combined UV exposure and V_{GS} bias effect on the a-IGZO TFT performance is rarely studied. There are studies showing that electrons, holes, and metastable defects in various parts of the a-IGZO layer are subject to the influence of the polarity and magnitude of the V_{GS} (15,18). Also, when the sample was illuminated under high energy UV, the lattice relaxation was difficult to recover (19). In this paper, authors study the phenomena of defects generation in the a-IGZO TFT under conditions of: UV exposure without the gate bias stress (UVS), with the positive gate bias stress (PBUVS), and with the negative gate bias stress (NBUVS), separately. The self-repair of these defects were investigated at room temperature and a high temperature.

Experimental

The bottom-gate a-IGZO TFT with an inverted staggered structure was fabricated on a glass substrate, as shown in Figure 1. After the 150 nm-thick molybdenum (Mo) film was sputter deposited and etched into the gate electrode, a 300 nm-thick SiO$_x$ gate dielectric layer was deposited by plasma-enhanced chemical vapor deposition (PECVD). Then, a 50 nm-thick a-IGZO semiconductor layer was sputtered deposited from an In$_2$O$_3$:Ga$_2$O$_3$:ZnO (1:1:1 mole ratio) target on the SiO$_x$ film at room temperature. Subsequently, a SiO$_x$ (200 nm-thick)/AlO$_x$ (20 nm-thick) stack was deposited as the etch stopper (ES) layer. After the source and drain contact vias were opened through the ES layer, a Ti/Al/Ti stack was deposited and etched into source and drain electrodes. Finally, a PECVD SiO$_x$ (100 nm-thick)/SiN$_x$ (100 nm-thick) stack was deposited as the passivation layer. The channel length (L) and width (W) of the TFT were 40 µm and 50 µm, respectively. Transfer characteristics of the TFT were measured at room temperature using an Agilent 4155C precision semiconductor parameter analyzer. The illumination light was an UV LED source (wavelength λ 395 nm equivalent to band gap energy of 3.14 eV) at an intensity of 0.3 mW cm^{-2}. For the UVS measurement, the TFT was exposed to the UV light without applying the V_{GS}. For PBUVS and NBUVS measurements, the TFT was exposed to the UV light and simultaneously applied with a V_{GS} of +20 V (for the former) or -20 V (for the latter) while both source and drain electrodes were grounded. Transfer characteristics were measured at $V_{DS} = 0.1$ V every 5 minutes. For the defect recovery study, the TFT was first stressed under the UVS, NBUVS, or PBUVS condition for 30 minutes. After removing the illumination light and the stress V_{GS}, the TFT was stored in air at room temperature or heated to 200°C. Then, transfer characteristics were measured. The μ_{eff} was extracted from the transconductance (g_m) vs. V_{GS} curve in the linear region using the following equation:

$$\mu_{eff} = \frac{L g_m}{W C_i V_{DS}} \qquad (1)$$

where C_i is the gate capacitance, and V_{DS} is the drain voltage. The V_{th} was defined as the V_{GS} at the drain current (I_{DS}) of 10 pA and the S is the change of V_{GS} when I_{DS} was increased from 10 to 100 pA.

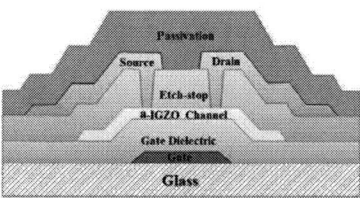

Figure 1. Cross-sectional view of the a-IGZO TFT.

Results and Discussion

Figure 2 shows transfer characteristics of an a-IGZO TFT measured in dark and under UVS. The inset table lists the μ_{eff}, V_{th}, and S corresponding to each curve. Compared with the TFT measured in dark, the UV illumination did not change μ_{eff} or I_{off}. However, it deteriorated other properties, i.e., a negative shift of V_{th} by 0.93V and a positive increase of S by 0.3 V/dec. The UV exposure can influence properties of several major parts of the TFT structure including 1) the ES layer, 2) the (ES/a-IGZO) back channel interface, 3) the bulk a-IGZO layer, 4) the (a-IGZO/gate dielectric) interface, and 5) the bulk gate dielectric layer. UV damages to the ES and gate dielectric layers can be neglected because they are high quality dielectric films, e.g., lack of hysteresis phenomena at difference temperatures (11). The independence of the μ_{eff} from the UV light illumination indicates the negligible amount of increase of free carriers in the a-IGZO layer. Since the band gap energy of the 395 nm UV is smaller than that of the a-IGZO, i.e., 3.14 eV vs. 3.3 eV, free carriers cannot be generated from the band-to-band excitation process. It is also very difficult to generate free charge carriers from the trap-assisted mechanism because traps would be located in the high density of states (DOS) tail regions close to the conduction or valence bands (20). Separately, the low I_{off} under the UV light can be explained by the reduction of oxygen vacancies at the ES/a-IGZO interface, which reduced the band bending (21-23). This is different from the phenomenon of the high I_{off} under the blue or green illumination condition, which generated free carriers from the trap-assisted process (24-26). These trap states are located in the mid-gaps, i.e., 2.34 eV -2.64 eV above the valence band or below the conduction band of the a-IGZO (11). In addition, the negative shift of the V_{th} in Fig. 2 can be contributed by the migration of positively charged V_O^{1+} and V_O^{2+} oxygen vacancies that were generated from the neutral oxygen vacancies V_O during the UV illumination of the a-IGZO film (23,25,26,27). The increase of S is caused by the increase of the gate dielectric interface defects, which interferences the channel formation process.

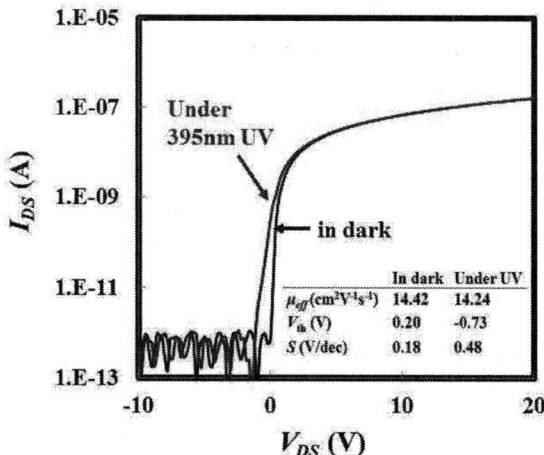

Figure 2. Transfer characteristics of a TFT in dark and under UVS.

In order to further understand the time dependence of the UV illumination effect, transfer characteristics of the a-IGZO TFT in dark and under various stress conditions were measured for an extended period of time. Shifts of S and V_{th} with respect to those of the original unstressed TFT measured in dark are expressed as ΔS and ΔV_{th}. Figures 3(a) and (b) show changes of ΔS and ΔV_{th} with time under UVS and PBUVS at $V_{GS} = 20V$ as well as NBUVS at $V_{GS} = -20V$, separately. Fig. 3 (a) shows that under the UVS or PBUVS condition the ΔS increased by 0.30 V/dec within a very short period of time. Then, it remained the same for 30 minutes. Therefore, the number of defects at the gate dielectric interface increased and saturated almost immediately upon the UV illumination. The application of the positive V_{GS} had little effect on the interface defects because a negligible amount of free electrons could be inducted from the a-IGZO layer to the gate dielectric interface. The NBUVS result is similar to that of the UVS or PBUVS, i.e., the ΔS increased and saturated almost immediately upon the stress. However, the magnitude of the former is consistently smaller than that of the latter, i.e., 0.19 V/dec vs. 0.30 V/dec. In principle, the negative V_{GS} bias could induce free positive holes to the gate dielectric interface (26). However, when defects at the gate dielectric interface were saturated, there is low chance that more free positively charged carriers could be induced to generate new defects. Therefore, the interface damage of the NBUVS is less serious than that of the PBUVS.

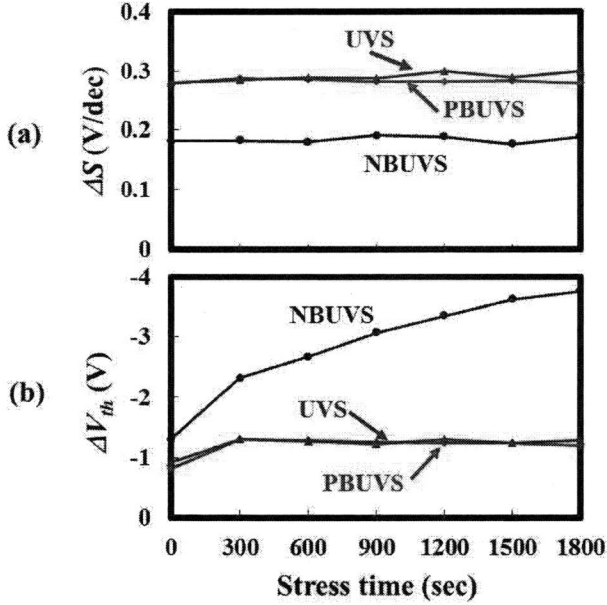

Figure 3. (a) ΔS and (b) ΔV_{th} vs. stress time under NBUVS, PBUVS and UVS.

Fig. 3(b) shows that upon the illumination of UV, the magnitude of the ΔV_{th} increased almost immediately and reached the saturation value within a few minutes. The saturation of the ΔV_{th} means the generation of V_O^{1+} and V_O^{2+} vacancies was maximized quickly upon the UV illumination. The lack of the influence of the positive V_{GS} on the ΔV_{th} is consistent with the previous discussion that the amount of free electrons in the UV illuminated a-

IGZO layer was too small to influence the amount of positive charge carriers at the gate dielectric interface. On the other hand, for the NBUVS condition, the magnitude of the ΔV_{th} increased throughout the 30-minute stress time without reaching a saturation value. For the same stress time, the magnitude of the ΔV_{th} of the NBUVS sample was always larger than that of the PBUVS sample. It was reported that the V_{th} shift could be contributed by the hole-trapping not only at the interface but also in the bulk of the gate dielectric layer (16,28). Therefore, under the NBUVS condition, positive charges in the UV illuminated a-IGZO layer were constantly induced to the bulk gate dielectric layer.

The time dependent self-repair of the UV induced damages in the a-IGZO TFT was studied. After being stressed under the UVS, PBUVS, or NBUVS condition at room temperature for 30 minutes, the TFT was stored in dark at room temperature or heated to 200°C and transfer characteristics were measured. Figure 4 shows curves of ΔS vs. relaxation time, i.e., time after releasing the stress condition. At room temperature, for the

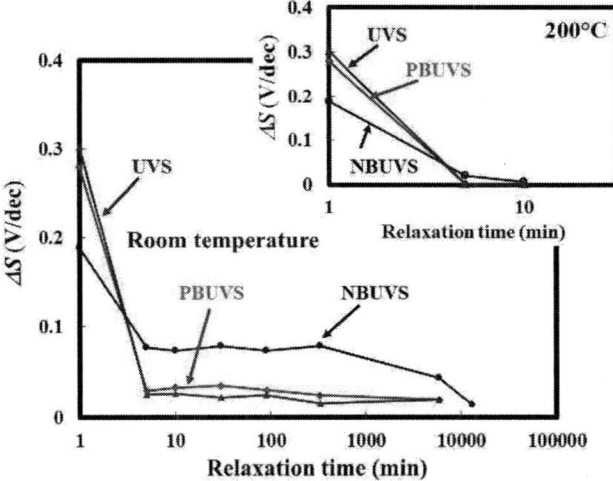

Figure 4. After NBUVS, PBUVS and UVS stresses ΔS vs. relax time at room temperature. Inset: at 200°C.

UVS stressed TFT, the ΔS dropped by 92% within 5 minutes. Subsequently, it was slowly reduced by about 4% in the next 5,880 minutes. For the PBUVS stressed TFT, the ΔS vs. time curve overlaps with that of the UVS stressed TFT. Therefore, defects on the gate dielectric interface were of two types, i.e., quickly or slowly self-repaired. It is possible that the former was composed of loosely trapped positive charges and the latter was neutral type dangling bonds. Separately, for the NBUVS stressed TFT, defects were recovered in three stages. In the first 5 minutes, they were reduced by 59%. In the next 330 minutes, they were slightly reduced. After 13,080 minutes, they were completely removed. Although at the beginning, the ΔS of the NBUVS TFT was smaller than that of the UVS or PBUVS TFT, i.e., 019 V/dec vs. 0.30 V/dec, the defect recovery rate of the former was much slower than that of the latter. Defects created from NVUBS may include: loosely trapped positive charge carriers at the gate dielectric surface, dangling bonds, and charges trapped in the bulk gate dielectric. They were self-repaired at difference rates. The defect repair rate was accelerated at the high temperature. The inset of Fig. 4 shows ΔS vs.

relaxation time curves at 200°C. For the UVS or PBUVS stressed TFT, it took less than 5 minutes to restore S to that of the original unstressed TFT. For the NBUVS stressed TFT, the recovery of S occurred in 2 stages. Defects in the bulk gate dielectric required more time to remove than those at the a-IGZO/gate dielectric interface did.

Figure 5 shows ΔV_{th} vs. relaxation time curves at room temperature. The inset shows curves measured at 200°C. At room temperature, for the UVS stressed TFT, the ΔV_{th} dropped by more than 50% within 5 minutes. It remained almost the same for the next 330 minutes and then reduced to about 40% after 5,880 minutes. The same trend was observed for the PBUVS stressed TFT. The ΔV_{th} is mainly related to defects generated in the bulk a-IGZO layer, e.g., V_O^{1+} and V_O^{2+} vacancies as well as dangling bonds. They were recovered at different rates. Their recovery rates are lower than those trapped at the gate dielectric interface. Fig. 5 also shows that the magnitude of the ΔV_{th} of the NBUVS TFT was consistently larger than that of the UVS or PBUVS TFT. Under the UV illumination, the transfer of positive oxygen vacancies could generate more defects in the a-IGZO layer than the transfer of electrons did. The former might take longer time to self-repair than the latter did. The restoring of ΔV_{th} was very sensitive to temperature. The inset of Fig. 5 shows that the V_{th} was restored to the original unstressed value within 10 minutes. The high thermal energy accelerated the reconnection of broken bonds and neutralization of charges in the a-IGZO layer.

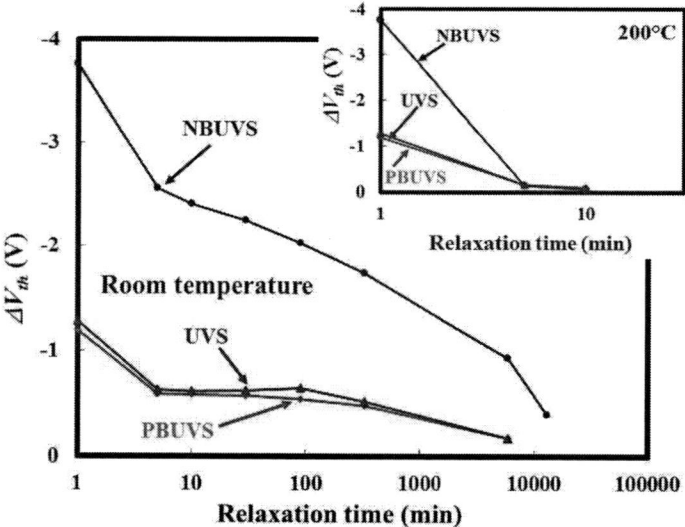

Figure 5. After NBUVS, PBUVS and UVS stresses ΔV_{th} vs. relax time at room temperature. Inset: at 200°C.

Conclusion

Defects generated in the a-IGZO TFT in dark and under UVS, PBUVS, and NBUVS stress conditions were investigated. Compared with initial properties of the TFT measured in dark, the UV illumination with or without the gate bias caused the negative shift of the

V_{th} and the increase of S. Defects were created and saturated at the gate dielectric interface as well as in the bulk a-IGZO layer within a short period of UV exposure. The addition of the positive gate bias voltage did not affect the mechanism of the defect-generation. Majority of these defects were loosely trapped and could be self-repaired after being stored in dark at room temperature for 5 minutes. On the other hand, with the addition of a negative gate bias voltage, defects were generated at the interface and in the bulk of the gate dielectric layer. The former saturated quickly but the latter increased with the stress time. Defects generated by the NBUVS were more difficult to self-repair than those generated by UVS or PBUVS. All defects generated by UV exposure with or without the gate bias could be removed after a short period of 200°C heating. In summary, under the UV illumination condition, the polarity of the gate bias voltage determines the type of defects generated in the TFT. This study provides important information on defects generation and removal in the a-IGZO TFT. The result is critical to the real world application of this kind of TFT.

Acknowledgments

Authors acknowledge Prof. Ting-chang Chang, Department of Physics of National Sun Yat-sen University, Kaohsiung, for providing samples in this study. Jingxin Jiang thanks the Doctoral Scientific Research Foundation of Liaoning Province (project No. 201601156), National Natural Science Foundation of China (project No. 6187011861), and China Scholarship Council for proving financial support for her visit to Thin Film Nano and Microelectronics Laboratory, Texas A&M University through the Postdoctoral Research Program.

References

1. T. Kamiya, K. Nomura, and H. Hosono, Sci. Technol. *Adv. Mater.* **11**, 044305 (2010).
2. M. Furuta, T. Kawaharamura, T. Uchida, D. Wang, and M. Sanada, *J. Display Technol.* **10**, 94 (2014).
3. J.-Y. Kwon, D.-J. Lee, and K.-B. Kim, *Electron. Mater. Lett.* **7**, 1 (2011).
4. J. Jiang, D. Wang, T. Matsuda, M. Kimura, S. Liu, and M. Furuta, *J. Nano Res.* **46**, 93 (2017).
5. K. Nomura, H. Ohta, A. Takagi, T. Kamiya, M. Hirano, and H. Hosono, *Nature* **432**, 488 (2004).
6. H. Oh, S, Yoon, M. Ryu, C. Hwang, S. Yang, and S. Park, *Appl. Phys. Lett.* **97**, 183502 (2010).
7. J. Jiang and M. Furuta, *J. Nanosci. Nanotechno.* **18**, 5668 (2018).
8. M. Furuta, J. Jiang, M. P. Hung, T. Toda, D. Wang, and G. Tatsuoka, *ECS J. Solid State Sci. Technol.* **5**, Q88 (2016).
9. J. Jiang, T. Toda, M. P. Hung, D. Wang, and M. Furuta, *Appl. Phys. Express* **7**, 114103 (2014).
10. X. Huang, C. Wu, H. Lu, F. Ren, Q. Xu, H. Ou, R. Zhang, and Y. Zheng, *Appl. Phys. Lett.* **100**, 243505 (2012).
11. T. Chen, Y. Kuo, T. Chang, M. Chen and H. Chen, *J. Phys. D: Appl. Phys.* **50**, 42LT02 (2017).
12. H. Lu, X. Zhou, T. Liang, L. Zhang, and S. Zhang, *IEEE J. Electron Devices Soc.*

5, 504 (2017).

13. Y. Zhang, L. Qian, Z. Wu and X. Liu, *Material* **10**, 168 (2017).
14. T. C. Fung, C. S Chuang, K. Nomura, H. P. D. Shieh, H. Hosono, J. Kanicki, *J. Inf. Display* **9**, 21 (2008).
15. M. D. H. Chowdhury, P. Migliorato, and J. Jang, *Appl. Phys. Lett.* **97**, 173506 (2010).
16. P. Liu, T. P. Chen, X. D. Li, Z. Liu, J. I. Wong, Y. Liu, and K. C. Leong, *Appl. Phys. Lett.* **103**, 202110 (2013).
17. L.-F. Tang，H. Lu，F.-F. Ren，D. Zhou，R. Zhang, Y.-D. Zheng, X.-M. Huang, *Chin. Phys. Lett.* **33**, 038502 (2016).
18. J.-H. Shin, J.-S. Lee, C.-S. Hwang, S.-H. K. Park, W.-S. Cheong, M. Ryu, C.-W. Byun, J.-I. Lee, and H. Y. Chu, *ETRI* 31,62 (2009).
19. J. H. Kim, U. K. Kim, Y. J. Chung, and C. S. Hwang, *Appl. Phys. Lett.* **98**, 232102 (2011).
20. Y. Kang, H. Song, H.-H. Nahm, S. H. Jeon, Y. Cho, and S. Han, *Appl. Phys. Lett. Mat.* **2**, 032108 (2014).
21. Y. J. Tak, D. H. Yoon, S. Yoon, U. H. Choi, *ACS Appl. Mater. Interfaces* **6**, 6339 (2014).
22. H. Y. Chong, S. H. Lee, T. W. Kim, *J. Electrochem. Soc.* **159**, B771 (2012).
23. K. Takechi, M. Nakata, T. Eguchi, H. Yamaguchi, and S. Kaneko, *Jpn. J. Appl. Phys.* **48**, 010203 (2009).
24. J. K. Jeong, *J. Mater. Res.* **28**, 2071 (2013).
25. B. Ryu, H.-K. Noh, E.-A. Choi and K. J. Chang, *Appl. Phys. Lett.* **97**, 022108 (2010).
26. S. Kim, S. Kim, C. Kim, J. Park, I. Song, J. Jeon, S.-E. Ahn, J.-S. Park and J. K. Jeong, *Solid-State Electron.* **62**, 77 (2011).
27. Y. S. Rim, W. Jeong, B. D. Ahn, and H. J. Kim, *Appl. Phys. Lett.* **102**, 143503 (2013).
28. H. Oh, S. M. Yoon, M. K. Ryu, C. S. Hwang, S. Yang, and S. H. K. Park, *Appl. Phys. Lett.* **98**, 033504 (2011).

Device Scalability of InGaZnO TFTs for Next-generation Displays

Saeroonter Oh[a], Su Hyun Kim[a], Mingoo Kim[a], Sang Min Yu[a], Youngjoon Choi[a], Joon Seok Park[b], and Jun Hyung Lim[b]

[a] Department of Electrical and Electronic Engineering, Hanyang University, Gyeonggi-do 15588, Korea
[b] R&D Center, Samsung Display Inc., Yongin 17113, Korea

> Device scaling of thin-film transistors has only recently gained attention, as the need has grown for shorter channel lengths in hybrid integration applications as well as a performance enhancer in future displays. In this study, we study the device scalability of short-channel top-gate oxide TFTs down to 1 μm, with particular focus on the carrier profile at the channel edges. We propose a carrier profile extraction method by utilizing comprehensive simulation fitting of the effective channel length, threshold voltage roll-off to the measured and extracted values. We find that the effective channel length plays an important role at short-channel lengths, and the scalability can be improved by adjusting the process-dependent carrier concentration profile.

Introduction

Metal-oxide semiconductor thin-film transistors (TFTs) are used as the backplane for large-screen, high-resolution OLED and LCD displays (1, 2). Amorphous InGaZnO (IGZO) TFTs are widely used in commercialized TVs, monitors, tablets, and wearable displays for its good device performance, stability, and process controllability. Also, due to its extremely low off-current, IGZO TFTs are used for low refresh rate and low-power displays. Furthermore, IGZO TFTs have a low process temperature, enabling fabrication on various substrates (including flexible films), and also hybrid integration with other backplane technologies, such as LTPS or Si CMOS. By taking advantage of the low off-current and low process temperature, IGZO can be monolithically integrated in the back-end of Si CMOS for low-power applications, as well as embedded memories for long retention and non-volatile operation (3, 4). There are many incentives to shorten the channel length of IGZO TFTs. First, shorter channels can boost the current without incorporating high-mobility oxides, which usually suffer from reliability issues and small process margin. When gate drivers are integrated directly onto the panel, larger currents allow smaller device footprint, thinner bezel width, and less deadspace. Second, transistors with sub-micron pitch are desired to bridge the pixel gap between smartphone displays and microdisplays, and to be used for consumer AR/VR displays with > 1000 ppi. Third, reduced device length is necessary for fine-grained integration with nanoscale CMOS and high-density memory arrays. However, short-channel TFTs suffer from short-channel effects (SCE), which can be seen from the threshold voltage (V_t) roll-off characteristics where the V_t reduces for short channel lengths. Process and device variability is inevitable and a nominal length positioned at a short channel length could have large V_t variations and process tolerance worsens. Understanding and control of SCE is critical for designing and fabricating short-channel TFTs.

In this paper, we study the device scalability of short-channel IGZO TFTs down to 1 μm. The electrical channel length (L_{eff}) is not the same as the designed or physical gate length (L_g), and is determined by the carrier profile at the gate edges. L_{eff} ($= L_g - \Delta L$) is obtained by the transmission line method (TLM), or shift-and-ratio method. Since scalability is related to the carrier concentration profile, it is strongly dependent on the process, where oxygen and hydrogen content varies with process conditions, and profiles change during thermal treatments. Devices with higher carrier concentration generally result in larger ΔL and worse SCE. Device simulations using TCAD shows the dependence on overlap (L_{ov}), steepness of carrier Gaussian decay (x_c), channel carrier concentration (N_{ch}) on the L_{eff} and SCE. TCAD simulation results are matched with experimental measurements for verification of the simulation models and parameters used. We define a critical carrier concentration (N_{crit}) which defines the L_{eff}. Changing the carrier profile while maintaining N_{crit} results in approximately the same ΔL. This study shows that SCE may be controlled through shaping of the carrier concentration profile through optimization of the device structure and process conditions.

Effective Channel Length Extraction

Several types of device lengths can be defined within a device. The actual gate length (L_{gate}) that can be measured optically or with an electron microscope is usually shorter than the designed mask length (L_{mask}) due to the lithography and etch bias. The electrical characteristics of the device is inversely proportional to the effective channel length (L_{eff}), not L_{gate}. ΔL is the difference between L_{gate} and L_{eff} (if we were to use $L_{mask} - L_{eff} = \Delta L$, then ΔL would include the litho and etch bias). ΔL is process-dependent, particularly the process responsible for the high-conductivity region formation, such as gate insulator dry etch, plasma treatment, and dopant diffusion. ΔL becomes non-negligible at short-channel devices, and a large ΔL could cause severe SCE. ΔL is generally extracted using the transmission line method (TLM). In this section, we will compare two L_{eff} extraction methods: TLM and shift-and-ratio method.

The total resistance (R_{tot}) between source and drain electrodes consists of the source-drain resistance (R_{sd}) and the channel resistance, as shown in Equation [1]:

$$R_{tot} = \frac{V_{DS}}{I_D} = R_{sd} + \frac{L_{gate} - \Delta L}{\mu_n C_{ox} W (V_{GS} - V_t - m V_{DS}/2)} \qquad [1]$$

where R_{sd} is bias-independent and the channel resistance is proportional to $L_{eff} = L_{gate} - \Delta L$. If we assume mobility and R_{sd} to be the same across different device lengths and plot R_{tot} vs. L_{gate} at several gate biases, ideally the lines will meet at one point. The (x, y) coordinates of this point will become (ΔL, R_{sd}). At short channel lengths, the drain bias affects the channel potential to a larger extent causing the threshold voltage to shift negatively. Due to this V_t roll-off, we need to maintain the same gate overdrive ($V_{GS} - V_t$) for different length devices for the lines to meet at one point. Fig. 1 shows an example of ΔL extraction from TLM. TLM is easy to use since it only requires I-V data of several devices with different lengths. However, we found that the ΔL value can differ for different ranges of lengths chosen. For example, ΔL value using L_{gate} = 1~40 μm devices is smaller than that when using L_{gate} = 10~80 μm devices, by 0.8 μm. The assumption of the same mobility and R_{sd} across devices should also be checked.

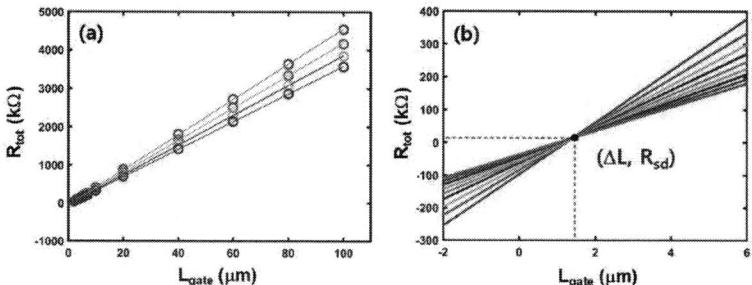

Figure 1. (a) R_{tot} vs L_{gate} lines of IGZO devices of various L_{gate} using TLM. (b) Linear extrapolation of R_{tot} vs L lines to shorter lengths to obtain ΔL.

Another method of extracting ΔL is the shift-and-ratio method (5), which may solve some of these issues. The main difference to TLM, is that there is no need to match the gate overdrive (does not use V_t), and also no assumption of the same R_{sd} across devices with different lengths. Following the work of Y. Taur et al. (5), R_{tot} is defined as:

$$R_{tot}(V_{GS}) = R_{sd} + L_{eff}\, f(V_{GS} - V_t) \tag{2}$$

where the channel resistance is expressed as a function of the gate overdrive and proportional to L_{eff}. After differentiating each side of Eq. [2], R_{sd} is removed and we obtain:

$$S(V_{GS}) \equiv \frac{dR_{tot}}{dV_{GS}} = L_{eff}\, \frac{df(V_{GS}-V_t)}{dV_{GS}} \tag{3}.$$

We can obtain the S value for a "long-channel" device and also for a "short-channel" device, which we denote as S_{long} and S_{short}, respectively. We found that extracted values are reliable when we use a "long-channel" that is more than 10 times the "short-channel". If V_t were to be the same, the S ratio (= r) would give us the effective channel length ratio, where we can easily solve for ΔL. Since V_t is different for short-channel devices, we incrementally shift S_{short} by δ, and find the δ ("shift") where the S ratio ("ratio") is constant with V_{GS}.

$$r \equiv \frac{S_{long}(V_{GS})}{S_{short}(V_{GS}-\delta)} \tag{4}.$$

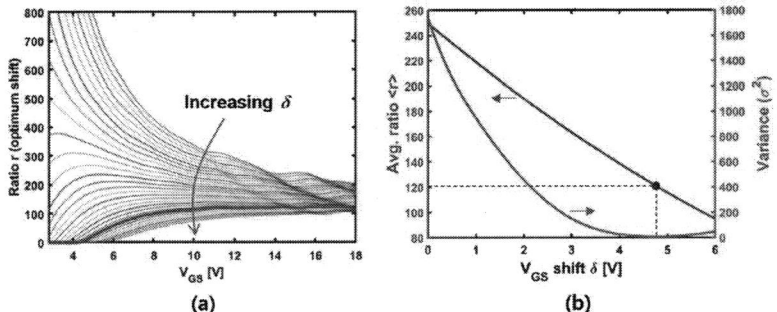

(a) (b)

Figure 2. (a) R_{tot} vs L_{gate} lines of devices of various L_{gate} using TLM. (b) Linear extrapolation of R_{tot} vs L lines to shorter lengths to obtain ΔL.

Fig. 2(a) shows the S ratio r with increasing V_{GS}-shift of S_{short}. Fig. 2(b) shows when the variance is the lowest (when r is the most constant with V_{GS}), and we denote the r value at that point as $r_{\delta min}$. Finally, ΔL is obtained by:

$$\Delta L = \frac{L_{short}\, r_{\delta min} - L_{long}}{r_{\delta min} - 1}$$ [5].

The shift-and-ratio method requires only two devices for ΔL extraction, unlike TLM which typically requires 5~10 device measurements. We checked to see if the choice of those two devices affect the resulting ΔL value. We extracted 29 ΔL values from combinations of short gate lengths of 1~6.5 µm and long lengths of 40~100 µm, and the average was 1.25 µm with little standard deviation of 0.12 µm, verifying that the shift-and-ratio method gives consistent ΔL values. Table 1 compares the ΔL values from TLM and shift-and-ratio methods, for three devices: device A (nominal), device B (higher carrier concentration), device C (lower carrier concentration). The drawback of the shift-and-ratio method is that noisy I-V data may give erroneous results since it uses differentials. Since both methods have its strengths and weaknesses, we selected the equally-weighted average of the two to be used in fitting TCAD simulations to measured results.

TABLE I. Comparison of ΔL extracted from TLM and shift-and-ratio.

Device	TLM (µm)	Shift-and-ratio (µm)	Average (µm)
A (nominal)	-0.18	-0.02	-0.10
B (higher carrier)	0.42	0.16	0.29
C (lower carrier)	-0.13	-0.32	-0.22

Carrier Profile and Short Channel Effects

Device Structure

Fig. 3(a) shows a schematic of the top-gate, top-contact device structure. IGZO channel is deposited on top of the SiO_2 buffer layer, and a SiO_2 gate insulator is used. High-conductivity source/drain extension regions are formed during the gate stack formation and via subsequent process steps. L_{eff}, which changes with the fabrication process, depends on the carrier profile and hence, has a large effect on the device scalability at short channel lengths. However, it is difficult to experimentally obtain the lateral carrier profile within the device. Therefore, we will use TCAD simulations, match the current-voltage (I-V) characteristics, and fit ΔL and the V_t roll-off to those of experimental measurements, by varying the carrier profile parameters. We propose a systematic method to obtain a unique combination of carrier profile parameters, and hence, the lateral carrier distribution profile.

Fig. 3(b) shows the main parameters used for the carrier profile in simulations. N_{ch} is the n-type carrier concentration of the channel; L_{ov} (overlap) is the position of the where the doping profile starts to decay (center of Gaussian distribution) in relation to the gate edge; x_c is the straggle (or standard deviation) of the Gaussian distribution. TCAD device simulations were performed by Silvaco ATLAS (6). The experimental and simulated device width is 7 μm, and gate length varies from 1 to 19.5 μm. Fig. 4 shows the measured and simulated initial I-V characteristics, where the device structural parameters, and density-of-states parameters were adjusted, so that the V_t, SS, mobility are in good agreement between simulations and experiments. Device A and B are the same devices from Table 1, which are the nominal and higher carrier concentration devices, respectively.

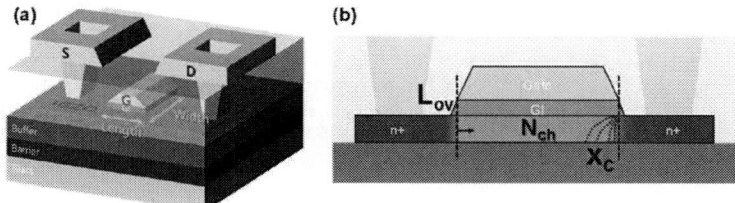

Figure 3. (a) Schematic of the device structure. The n+ regions denote the high-conductivity source/drain extension regions. (b) Parameters for expressing the channel carrier profile: L_{ov} (overlap in relation to gate edge), x_c (straggle of the Gaussian distribution), and N_{ch} (channel carrier concentration).

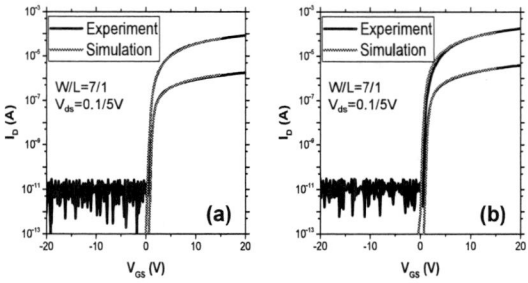

Figure 4. Experimentally measured (black lines) and simulated (red lines) I_D-V_{GS} characteristics of (a) device A and (b) device B. Device width and gate length are 7 μm and 1 μm, respectively. Curves at $V_{DS} = 0.1$ V and 5 V are shown.

Carrier Profile Extraction

We now consider the relation between ΔL and the carrier profile. First, we apply the measured ΔL and obtain the carrier concentration at both edges of L_{eff}. We will denote this carrier density as N_{crit}. We can also extract ΔL from simulated devices with varying L_{ov} and x_c parameters. Fig. 5(a) shows a constructed ΔL contour map by extracting the ΔL from $1 \sim 19.5$ μm gate lengths for each matrix combination of the L_{ov} and x_c values. For each carrier profile parameter combination and extracted ΔL, we can also define the N_{crit} from simulations. We found that for the same ΔL the N_{crit} is the same, even though the carrier profile is largely different. A larger L_{ov} with a steeper profile may give the same ΔL value with a device with a smaller (or negative) L_{ov} with a less steep profile. These profiles with the same ΔL meet at one point, and the carrier concentration at that point is N_{crit}. The ΔL values taken from N_{crit} (ΔL_{Ncrit}) matches fairly well with ΔL values obtained by TLM (ΔL_{TLM}). So just by taking the position of the N_{crit} point on an arbitrary carrier profile with relation to the gate edges, we can obtain ΔL.

What is the physical meaning of N_{crit}? From (7), N_{crit} is the position where the resistivity of the source/drain extension regions becomes the same as the channel resistivity. The n-doping concentration in the source/drain extension regions is a function of x (position) and falls going into the channel (resistivity increases). The channel carrier concentration is a function of x (position) and the bias. Since we already have information on ΔL and N_{crit}, we found that the channel carrier concentration at the ΔL point at $V_{GS} = 2.5$ V becomes N_{crit},

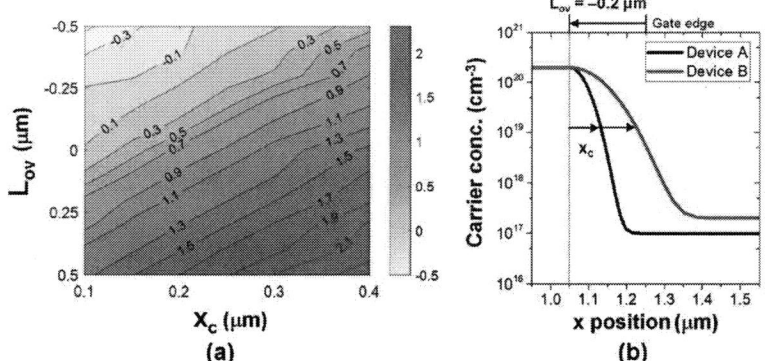

Figure 5. (a) ΔL contour map while varying L_{ov} and x_c. The color bar and contour lines show the ΔL values in μm. (b) Extracted lateral carrier profile of device A and B.

As can be seen in Fig. 5(a), ΔL increases as L_{ov} or x_c increases. Since a doping profile reaching further into the channel will aid the lateral electric fields to penetrate into the channel, SCE will exacerbate. Since a large ΔL means shorter L_{eff}, a small ΔL is required to keep the SCE under control. A negative ΔL means L_{eff} is larger than the physical L_{gate}, which occurs at large underlap ($L_{ov} < 0$) situations. $\Delta L < 0$ may be advantageous towards SCE, but there's a trade-off with higher series resistance.

For device A and B, the gate stack formation process was the same, hence, the same L_{ov} values were used. To match ΔL only one x_c value is obtained. Fig. 5(b) shows the extracted lateral carrier profile of devices A and B. The higher carrier concentration of device B is reflected in N_{ch} and the larger x_c. Especially the longer carrier profile tail into the channel results in larger ΔL and worse SCE for device B compared to device A. By fitting the simulated I-V, ΔL, and V_t roll-off with those of experimental values, we end up with one unique combination of carrier profile parameters. Table II lists the main carrier profile parameters that are extracted by using this method.

TABLE II. Carrier profile parameters extracted from TCAD simulations fitted to measurement data.

Parameters	Device A (nominal)	Device B (high carrier density)
L_{ov} (μm)	-0.2	-0.2
x_c (μm)	0.07	0.15
N_{ch} (cm^{-3})	1×10^{17}	2×10^{17}

Conclusion

We studied the device scalability of top-gate IGZO TFTs down to 1 μm gate length. The effective channel length, while having the same gate length, is determined by the process-dependent ΔL, which can be extracted by TLM or shift-and-ratio methods. We extracted the lateral carrier profile within the device by fitting simulated I-V and ΔL characteristics to experimental measurements. Particularly, we found that the lateral carrier profile overlap and straggle significantly affects the SCE, and ΔL can be found from N_{crit}. The higher carrier concentration device with worse SCE has a more gradual carrier distribution tail where the lateral fields can penetrate into the channel. Knowing and understanding the lateral carrier profile in relation to the process conditions will be essential in ensuring TFT scalability to sub-micrometer lengths.

Acknowledgments

This work is supported by Samsung Display Co., and in part by the research fund of Hanyang University (HY-2017-N). The EDA tool was supported by the IC Design Education Center (IDEC).

References

1. S. Oh, J.H. Baeck, J.U. Bae, K.-S. Park, and I.B. Kang, *Appl. Phys. Lett.* **108**, 141604 (2016).
2. S. Oh, J.H. Baeck, H.S. Shin, J.U. Bae, K.-S. Park, and I.B. Kang, *IEEE Electron Device Lett.*, **35**(10), p. 1037 (2014).
3. S.H. Wu, X.Y. Jia, X. Li, C.C. Shuai, H.C. Lin, M.C. Lu, T.H. Wu, M.Y. Liu, J.Y. Wu, D. Matsubayashi, K. Kato, and S. Yamazaki, in *Symp. VLSI Tech. Dig.*, p. T166 (2016).

4. H. Inoue, T. Matsuzaki, S. Nagatsuka, Y. Okazaki, T. Sasaki, K. Noda, D. Matsubayashi, T. Ishizu, T. Onuki, A. Isobe, Y. Shionoiri, K. Kato, T. Okada, and S. Yamazaki, *IEEE J. Solid-State Circuit.*, **47**(9), p. 2258 (2012).
5. Y. Taur, D.S. Zicherman, D.R. Lombardi, P.J. Restle, C.H. Hsu, H.I. Hanafi, M.R. Wordeman, B. Davari, G.G. Shahidi, *IEEE Electron Device Lett.*, **13**(5), p. 267 (1992).
6. *Atlas User's Manual*, version August 26, 2016, Silvaco, Inc., Santa Clara, CA.
7. Y. Taur, T.H. Ning, *Fundamentals of Modern VLSI Devices*, p. 211, Cambridge University Press (1998).

TCAD Simulation of a 3D NAND Memory Utilizing In-Ga-Zn-Oxide: "3D OS NAND" with 4 V Drive, High Endurance and Density

H. Kunitake, H. Kimura, K. Tsuda, H. Godo, T. Murakawa, H. Sawai, H. Baba, S. Sasagawa, T. Ikeda, and S. Yamazaki

Semiconductor Energy Laboratory Co., Ltd., 398 Hase, Atsugi, Kanagawa, 243-0036, Japan

> We have evaluated, through TCAD simulation, 3-dimensional oxide-semiconductor NAND (3D OS NAND), a novel vertically-stacked 2T-1C memory using cylindrical *c*-axis aligned crystalline In-Ga-Zn-Oxide (CAAC-IGZO) channels. Taking advantage of low off-leakage of CAAC-IGZO field-effect transistors (CAAC-IGZO FETs), the memory does not use a metal-oxide-nitride-oxide-silicon (MONOS) structure for writing. This memory architecture is potentially capable of achieving a memory density and a 10-year retention similar to NAND flash, operating with a supply voltage of 4 V or below, while overcoming the endurance challenge of NAND flash.

Introduction

Memory-centric computing has been actively researched for AI application (1). To achieve this, there are growing demands on memories for high density and high endurance.

However, Si-based memories have not been able to resolve the trade-off between latency and density. To overcome this challenge, a variety of emerging memory technologies have been proposed (2)-(4). Although emerging memories are expected to improve the latency, they still have difficulties in achieving a density comparable to NAND flash and in improving endurance.

C-axis aligned crystalline In-Ga-Zn-Oxide (CAAC-IGZO), with a wide bandgap (5) and a heavy hole effective mass, is characterized by its extremely-low off-leakage current (6), (7). Use of CAAC-IGZO in non-display fields has been reported in recent years, and 3D NAND utilizing IGZO has been researched (8). In addition, 3D NAND using solid-phase crystallization with nickel has been reported (9)-(16). However, the reported memory cell architecture involves through-dielectric charge trapping in MONOS structures, and has low endurance in the same way as conventional NAND flash.

In view of the above, we propose a 3-dimensional oxide-semiconductor NAND (3D OS NAND) memory, a vertically-stackable 2T-1C memory having CAAC-IGZO cylindrical active layers. Potentially, this memory not only enjoys main features of NAND flash such as high density and non-volatility, but also achieves a DRAM-like endurance. We evaluated our novel memory architecture with TCAD Sentaurus produced by Synopsys.

The Concept of "3D OS NAND"

Memory Cell Architecture

The memory cell is configured to take advantage of properties of CAAC-IGZO as the channel material. The cross section of the memory cell is shown in Fig. 1(a), and the circuit diagram corresponding to the cross section is shown in Fig. 1(b). The unit cell has a 2T-1C configuration including vertically-positioned multilayered circular transistors. Two types of transistors each having one active layer are placed on the X-axis; the transistors with their active layers connected to WBL (i.e., write transistors) send, hold and release charges in a storage capacitor between SN and CG in accordance with on/off of WG, and the transistors with their active layer connected to RBL (i.e., read transistors) adjust current flowing to RBL in accordance with both charges written to storage capacitors and voltages applied to CG. Charges at storage capacitors correspond to written data and CG voltage corresponds to read address selection in the height direction. These functions constitute a memory operation. We assume that the active layers of 3D OS NAND are each uniform in carrier concentration, and each of the ends makes a junction with metal.

Based on the cell configuration proposed above, our proposed memory cell architecture example of 3D OS NAND is shown in Fig. 1(c). In Fig. 1(c), NAND strings each with the architecture shown in Fig. 1(a) are placed with an assumed close-packing (circle packing) structure with 200 nm pitch (strings with a diameter of 180 nm and 20-nm spacing). The four metal layers, two placed on the top and two at the bottom, in Fig. 1(c) with gold color are bit lines (BLs) (RBL, WBL, BWBL, and BRBL from top to bottom), and are each separated into 4 lines in the Y-direction, which work as column selectors. The staircase consisting of 8 metal layers with alternate 2 colors are WGs or CGs. These WGs and CGs are shared in 16 strings in a NAND block and work as height selectors. In Fig. 1(c), 2 colors are used for easy distinction between WGs and CGs, not for representing different materials. The top metal layer with violet color composed of 4 lines in the X-direction also represents WGs. In this proposal these top WGs work as row selectors as well as height selectors for top cells.

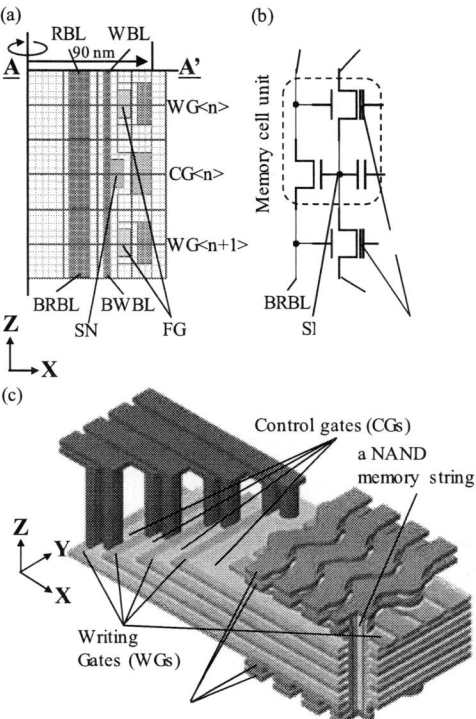

Fig. 1. (a) A cross section of a NAND memory string and (b) equivalent circuit diagram of 1.5 unit cells of 3D OS NAND discussed in this paper. (c) A 3D image of 3D OS NAND block.

Such an operation of the 3D OS NAND is possible because the off-state leakage current of the transistor with a CAAC-IGZO active layer is very low (6), which enables a long retention time even when the channel of the transistor and the storage node are electrically connected. Thus, unlike NAND flash, the 3D OS NAND does not require charge injection through dielectric films, which leads to high endurance and a low drive voltage. In addition, since CAAC-IGZO films can be deposited at 400°C or lower (17) or by an ALD process (18), setup of the process for this memory technology is estimated to be relatively simple. These properties also allow the memory to be stacked over Si without substantial difficulties, which opens its applicability to memory-centric computing with peripheral circuits being placed below the memory cells (19). On the other hand, the complex cell configuration of 3D OS NANDs decreases the spatial cell density. The increase in the number of layers in a NAND string enlarges the string radius in the X-Y plane and the increase in device numbers per cell enlarges the cell pitches in the Z-direction. Although the 3D OS NAND inevitably suffers a disadvantage against conventional NAND flash in terms of spatial cell density, 3D OS NANDs can have the advantage of TBW (total byte written) density resulting from much higher rewrite endurance.

Memory Operation

Shown in Fig. 2 is a timing chart of write and read operations of a 3D OS NAND string. The timing chart illustrates a case where four memory cells are connected to a string. In write, RBL voltage is written to all the cells positioned on the drain terminal (WBL) side of the target cell. Thus, write operation needs to be performed from the cell furthest from RBL. In addition, 4 V voltage is applied on not only WGs but also CGs on WBL side of the target cell. This is because each active layer of 3D OS NAND in this work has uniform carrier concentration as mentioned above, field effect of CGs is also necessary for higher write current. The read operation can follow the read operation of NAND flash.

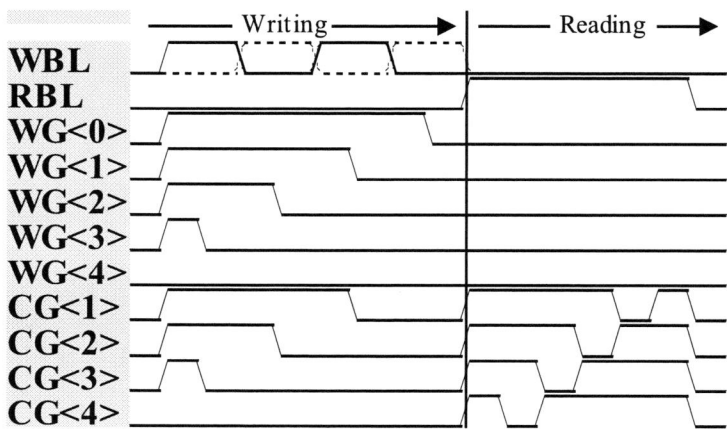

Fig. 2. A timing chart of write and read operations of a 3D OS NAND string.

Rewrites in 3D OS NANDs do not involve block erase operation by hole injection as in NAND flash. This is because cell writes are performed with current through the channel of the write FET; thus, writing the data of "0" or "1" makes no difference other than WBL voltages in 3D OS NANDs. If all the data in a block are to be erased, the erasure is as simple as writing "0" into all cells in the block. On the other hand, one disadvantage exists in rewrite operation in 3D OS NAND. As mentioned above, same data is written to all the cells positioned on the drain terminal (RBL) side of the target cell. So data that should not be rewritten need to be stored in back up and be rewritten after rewriting into the target cell.

The challenge of the above memory cell architecture is that writing data through a transistor channel leads to a shorter data retention time when compared with a MONOS structure. To improve retention, we propose a method for positively shifting the threshold voltage of the write FET. Specifically, a floating gate (FG) made of acceptor boron-doped poly Si is provided in the write FET of the 3D OS NAND, and electron injection is carried out as in NAND flash. The data storage mechanism in NAND flash is applied to the 3D OS NAND of this work in order to control threshold voltage of write FETs. The threshold voltage control is achieved through providing boron-doped poly Si FGs and injecting electrons into them. Though there are other charge storage methods such as charge trapping layers made of silicon nitride, threshold voltage of the write FET can be sufficiently controlled with this method. This method will enable a memory with configuration of write transistors and storage capacitors having a long retention time and limitless endurance with one-time electron injection.

Device Simulation

Simulation Parameters

Structure parameters of a 3D OS NAND which were used for the simulations are listed in TABLE I, and material parameters of CAAC-IGZO are listed in TABLE II. The radius of a 3D OS NAND string is 90 nm, and half-cell pitches for both write and read segments are 50 nm (the total pitch of 100 nm per unit cell.) FG material is the boron-doped poly Si and the other metal layers (WGs, CGs, and electrodes of storage capacitors) are tungsten. The interlayer insulator material between WGs and CGs is silicon oxynitride, and the other insulator materials are ordinary silicon oxide.

The electron affinity, band gap, density of state, and electron mobility of CAAC-IGZO are based on the experimental results of a single film (6)-(7), (20). The other parameters of IGZO are from literature in principle (21). Degradation models such as BTI are not considered in this simulation.

Some recent research report an IGZO electron mobility of 50-60 cm2/Vs (22)-(23). These values are evaluated from planar devices having μm-order dimensions. In this simulation, we used experimental results obtained from a single CAAC-IGZO film (6). The supply voltages used in the device simulation of write/read operations are listed in TABLE III. Although 15 V is used, it needs to be applied only once by the manufacturer to inject charge into the floating gate for threshold voltage adjustment of the write FET. The supply voltage for write/read in normal operation in an end user's product would be 4 V and does not require a high voltage during the memory operation, unlike in NAND flash.

Note that the TCAD simulations discussed here were run using a test device design made up of eight vertically positioned cells.

TABLE I
STRUCTURE PARAMETERS OF A 3D OS NAND.

position		parameter	value	unit
half cell pitch		Writing	50	nm
		Reading	50	nm
Inner FET (Reading line)	Structure	Gate length	30	nm
	Center filler	Permittivity	3.9	
		Radius	30	nm
	CAAC-IGZO	Thickness	15	nm
		Carrier density	1×10^{18}	cm^{-3}
	TGI	Permittivity	3.9	
		Thickness	10	nm
Capacitor	Storage node Electrode	Work function	4.7	eV
		Thickness	10	nm
	Insulator	Permittivity	3.9	
		Thickness	5	nm
	Control gate electrode	Work function	4.7	eV
Outer FET (Writing line)	Structure	Gate length	30	nm
	CAAC-IGZO	Thickness	5	nm
		Carrier density	1×10^{16}	cm^{-3}
	Tunneling insulator	Permittivity	3.9	
		Thickness	5	nm
	Floating gate	Material	Poly Si	
		Dopant	Boron	
		Doping density	1×10^{20}	cm^{-3}
		Permittivity	20	nm
		Thickness	10	nm
	TGI	Permittivity	3.9	
		Thickness	5	nm
	Writing gate electrode	Work function	4.7	eV

TABLE II
MATERIAL PARAMETERS OF CAAC-IGZO

	parameter	value	unit
conduction band	Electron affinity	4.6	eV
	Effective DOS	5×10^{18}	cm^{-3}
	Band-tail	Not defined.	
	Electron mobility	8	cm^2 / Vs
valence band	Band gap	3.0	eV
	Effective DOS	5×10^{18}	cm^{-3}
	Band-tail	Not defined.	
	Hole mobility	0.01	cm^2 / Vs
trap band	-	Not defined.	
	Relative permittivity	15	

TABLE III
VOLTAGE IN SEQUENTIAL WRITE/READ OPERATION SIMULATION OF AN 8-LAYER 3D OS NAND STRING.

Electrode	operation	voltage (V)
WG(n)	Pre-charge (Only one time operation before shipment is expected.)	0 - 15
	Write/Read	0 - 4
CG(n)	Write/Read	0 - 4
WBL	Write/Read	0 - 4
RBL	Write/Read	0 - 1.2
BWBL, BRBL	Write/Read	0

TCAD Simulation Results

The I_d-V_{wg} curves and the threshold voltage of the write FET are shown in Fig. 3. In this paper we focus on the retention time, which is determined solely by the threshold voltage and subthreshold swing of the write FET. That is because only subthreshold leakage is the main leakage factor of the CAAC-IGZO FET that affects the retention time of the memory using CAAC-IGZO FETs (6). The threshold voltage of the write FET shifts positively with higher pre-charge voltages (V_{prep}). V_{prep} is applied during charge injection into the floating gate of the write FET, determining its threshold voltage.

Shown in Fig. 4 are the retention properties of 3D OS NAND under the conditions of V_{prep} = 12.5 V or 15 V. In evaluating the retention properties, two different data patterns, a checker pattern and a reversed checker pattern, were written. The currents considered in this simulation are the following two; channel current between neighboring cells and direct tunneling current through an insulator film. As shown in Fig. 4, in order for the device to achieve 10-year (3.2×10^8 s) retention, application of $V_{prep} = 15$ V, i.e., a write-FET threshold voltage of around 2 V is required. The following simulations are run with the memory cells with charge injections with $V_{prep} = 15$ V and thus capable of 10-year retention.

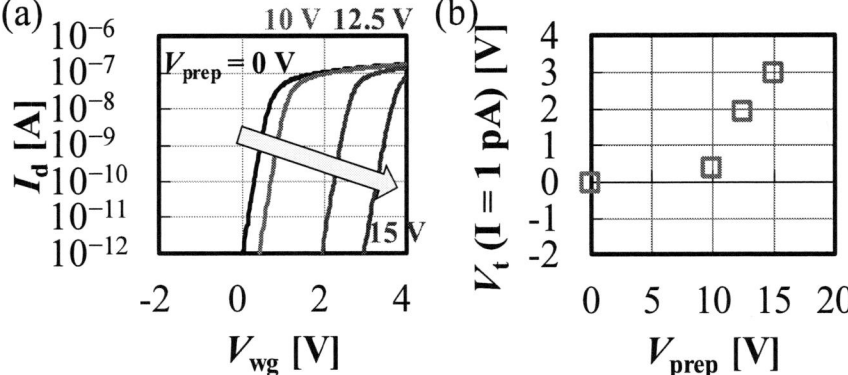

Fig. 2. (a) I_d-V_{wg} curves and (b) threshold voltage of a write FET in a 3D OS NAND string under various pre-charge voltage (V_{prep}).

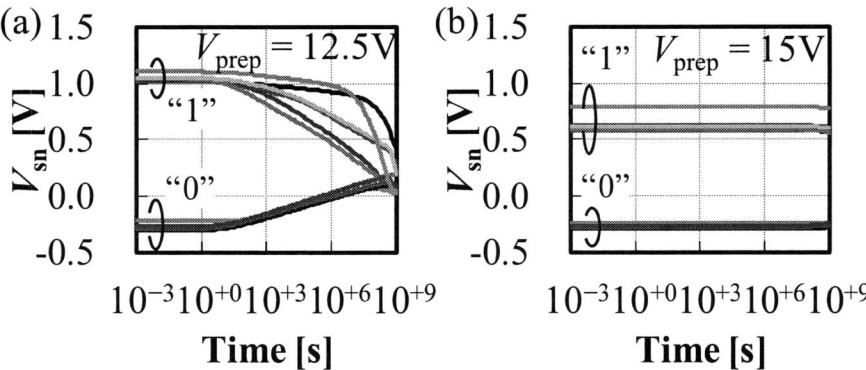

Fig. 3. Retention curves of storage nodes in a 3D OS NAND under a pre-charge voltage (V_{prep}) of (a) 12.5 V or (b) 15 V, with an input voltage of 0 V.

The simulation results of successive write and read operations are shown in Figs. 5(a) and 5(b). The voltage changes at the storage nodes at the time of write operation regarding two cells (SN<7>, SN<6>) at the bottom of a NAND string are shown in Fig. 5(a). We assume that a longer distance from WBL results in longer write time; focusing on the deepest cells enables examination of the worst case. In Fig. 5(a), response to writes of two data patterns, a checker pattern and a reversed checker pattern, is shown.

VSN window, which is defined as the VSN difference between two data patterns (checker and reversed checker), is lower than ordinary NAND flash, in which 4 V or higher voltage window is achieved. Some reasons for this phenomenon may be due to lower operation voltage, or voltage window losses caused by a high threshold voltage of write FETs, for example. Lower voltage window due to lower operation voltage can be an advantage, because it leads to lower power consumption. Low-voltage operation enables low power, and the fact that the device can operate at a low voltage also may enable multi-level cells if the voltage is increased. In contrast, the voltage window loss caused by high threshold voltage results in lower accuracy and reliability, and is to be suppressed in the future.

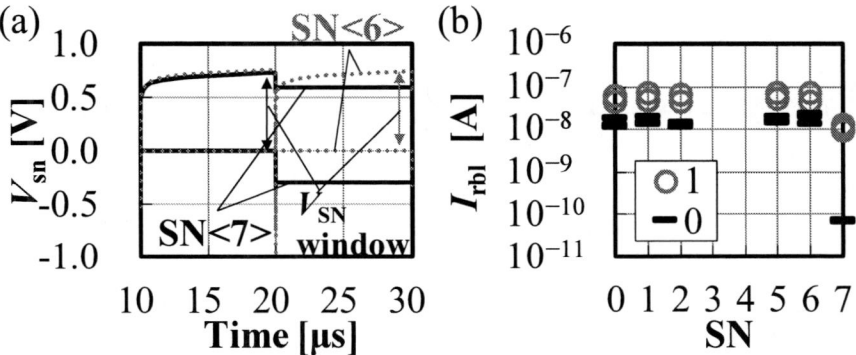

Fig. 4. (a) Voltage change of storage node 7 (SN<7>) and 6 (SN<6>) in writing operation of checker pattern. (b) Reading current of each storage node with various data patterns.

It is reported that with continuous positive bias into gate electrode, the threshold voltage of CAAC-IGZO FET shows positive drifts, which can partly be recovered after the device is turned off for a certain period (called BTI degradation)[24], [25], [26]. Although this phenomenon is not considered in this paper, this may decrease the read current, and lead to read failure in the worst case. Modeling of degradation of IGZO FETs and the degradation compensation to circuit operations would be topics for a future work. In Fig. 5(b), read current at each node in response to the inputs of "0" and "1" is shown. The read performance is similar at every node excluding SN<7>. SN<7> can be ignored by being used as dummy layers.

In the 3D OS NAND with this architecture, field effect due to the voltage increase of the RBL during read induces carriers of the WBL, possibly resulting in leakage current through the channel of write FET. This is known as read disturb. During read operation, there are two types of cells that are under two different bias conditions: one is a waiting cell having 4 V applied to its CG and thus in an on state, the other is a selected cell having 0 V applied to its CG and whose on/off depends on the data written in the storage capacitor. Fig. 6 shows retention curves in the conditions where voltage application during data read is reproduced. Fig. 6 (a) shows the retention curve of waiting cells with the bias of $V_{rbl} = 1.2$ V, $V_{cg} = 4$ V and Fig. 6 (b) selected cells with $V_{rbl} = 1.2$ V, $V_{cg} = 0$ V. In waiting cells, voltages at the storage nodes (V_{sn}) increase because of capacitive coupling with their respective CGs receiving a 4 V bias. Only at SN<0>, which is the node nearest to RBL, V_{sn} begins to drop down to 0 V (= Vwbl) from t = 10^4 s independent of written data. In contrast, in selected cells, the memory window of all cells appears to be slightly narrower at t = 10^9 s, as compared with Fig. 4(b). These data indicate the occurrence of read disturb. Among waiting cells, the cell nearest to RBL suffers from read disturb because of the field effect from RBL to which a positive bias is applied, and its retention time decreases to 10^4 s.

There are some possible solutions for compensation of this phenomenon. The simplest solution is to regard the cell nearest to RBL as the dummy cell. Another solution is refresh. Determining the allowable cumulative number of read operations and data rewrites after every certain number of read operations can avoid read disturb. As an example, if the page read time is assumed as 1 ms, the allowable cumulative number of page read is estimated as 10^7 times. In addition, selected cells achieve 10-year retention even with the effect of read disturb. Even in the case where the retention time of the cell during cumulative read operations is shorter than 10 years because of read disturb, the cell can be refreshed as described above.

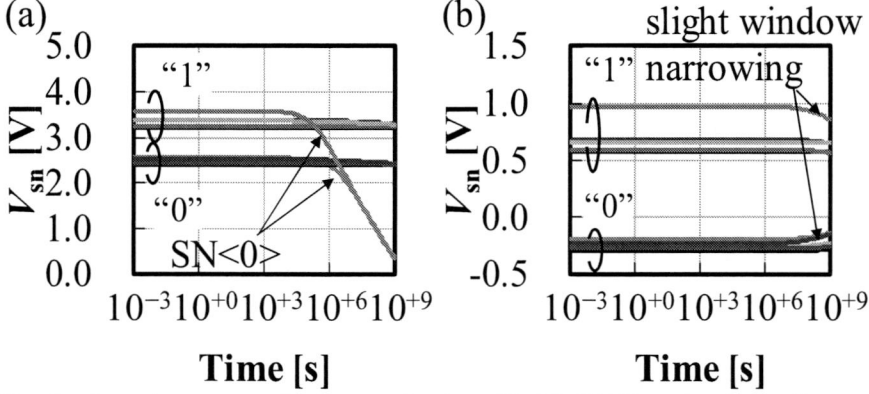

Fig. 5. Retention curves of storage nodes in a 3D OS NAND under the reading-out condition in (a) waiting cells ($V_{rbl} = 1.2$ V, $V_{cg} = 4.0$ V) and (b) selected cells ($V_{rbl} = 1.2$ V, $V_{cg} = 0.0$ V).

Conclusion

We proposed a novel memory architecture, 3D OS NAND with CAAC-IGZO active layers, and examined its properties through device simulation. The comparison between the 3D OS NAND, DRAM, and NAND flash is shown in TABLE IV.

Our proposed 3D OS NAND is potentially capable of achieving 10-year retention and reducing supply voltage. Since writing to 3D OS NAND cell is through a channel of a transistor, its endurance should be comparable to that of DRAM. Our 3D OS NAND shows a slightly lower bit density than NAND flash due to its configuration, but has capability for larger TBW density with its large endurance advantages. This feature may enable near-memory training in addition to near-memory inference utilizing 3D-OS-NAND-embedded AI accelerators.

TABLE IV
COMPARISON TABLE

Evaluation	DRAM	3D NAND Flash	3D OS NAND
Operating voltage	3.3 V	12 V	4 V
Endurance	∞	10^4	∞
Unit area (μm^2)	0.00462 (27)	0.0285 (28)	0.0346 (assuming the string pitch as 200 nm)
Density (bit/μm^2) (assuming 16 cells per string)	216	561	412
Retention	60 ms	10 year	10 year
Selector/peripheral	Complex	Simple	Simple

References

1. E. Beigne, "Emerging Technologies for Memory-centric and Low-power Architectures," *IEEE IEDM*, SC2, 2019.
2. G. W. Burr, R. M. Shelby, S. Sidler, C. d. Nolfo, J. Jang, I. Boybat, R. S. Shenoy, P. Narayanan, K. Virwani, E. U. Giacometti, B. N. Kurdi, and H. Hwang, "Experimental Demonstration and Tolerancing of a Large-Scale Neural Network (165 000 Synapses) Using Phase-Change Memory as the Synaptic Weight Element," *IEEE T-ED*, Vol. 62, No.11, 2015. 10.1109/TED.2015.2439635
3. K. C. Chun, H. Zhao, J. D. Harms, T. Kim, J. Wang, and C. H. Kim, "A Scaling Roadmap and Performance Evaluation of In-Plane and Perpendicular MTJ Based STT-MRAMs for High-Density Cache Memory," *IEEE JSSC*, vol. 48, no. 2, 2013. 10.1109/JSSC.2012.2224256
4. A. Kawahara, R. Azuma, Y. Ikeda, K. Kawai, Y. Katoh, Y. Hayakawa, K. Tsuji, S. Yoneda, A. Himeno, K. Shimakawa, T. Takagi, T. Mikawa, and K. Aono, "An 8 Mb Multi-Layered Cross-Point ReRAM Macro With 443 MB/s Write Throughput," *IEEE JSSC*, Vol. 48, No. 1, 2013. 10.1109/JSSC.2012.2215121
5. D. Matsubayashi, Y. Asami, Y. Okazaki, M. Kurata, S. Sasagawa, S. Okamoto, Y. Iikubo, T. Sato, Y. Yakubo, R. Honda, M. Tsubuku, M. Fujita, T. Takeuchi, Y. Yamamoto, and S. Yamazaki, "20-nm-node trench-gate-self-aligned crystalline In-Ga-Zn-Oxide FET with high frequency and low off-state current," *IEEE IEDM*, pp. 6.5.1-6.5.4, 2015. 10.1109/IEDM.2015.7409641
6. H. Kunitake, K. Ohshima, K. Tsuda, N. Matsumoto, T. Koshida, S. Ohshita, H. Sawai, Y. Yanagisawa, S. Saga, R. Arasawa, T. Seki, R. Honda, H. Baba, D. Shimada, H. Kimura, R. Tokumaru, T. Atsumi, K. Kato, and S. Yamazaki, "A c-Axis-Aligned Crystalline In-Ga-Zn Oxide FET With a Gate Length of 21 nm Suitable for Memory Applications," *IEEE J-EDS*, vol. 7, pp. 495-502, 2019. 10.1109/JEDS.2019.2909751
7. S. Yamazaki and M. Fujita, "Physics and Technology of Crystalline Oxide Semiconductor CAAC-IGZO Application to LSI," *WILEY*, 2017.

8. S. Choi, B. Kim, J. K. Jeong, and Y. H. Song, "A Novel Structure for Improving Erase Performance of Vertical Channel NAND Flash with an Indium-Gallium-Zinc-Oxide Channel," *IEEE T-ED*, Vol. 66, No. 11, 2019. 10.1109/TED.2019.2942935
9. S. Yamazaki, "Continous Grain Silicon Technology and Its Applications for to System on Panel," AM-LCD, 1998.
10. Y. Hirakata, M. Sakakura, S. Eguchi, Y. Shionoiri, S. Yamazaki, H. Washio, Y. Kubota, N. Makita, "4-in. VGA Reflection-Type Poly-Si TFT LCD with Integrated Digital Driver Using Seven-Mask CG Silicon CMOS Process," SID, 2000.
11. T. Takayama, H. Ohtani, A. Miyanaga, T. Mitsuki, H. Ohnuma, S. Nakajima, S. Yamazaki, "Continous Grain Silicon Technology and Its Applications for Active Matrix Display," AM-LCD, 2000.
12. S. Yamazaki, H. Ohnuma, K. Dairiki, T. Mitsuki, T. Takayama K. Akimoto, Method for manufacturing semiconductor device, JP Patent 4176362, 2001-03-16.
13. K. Kato, A. Isobe, H. Miyairi, S. Yamazaki, Semiconductor memory device, JP Patent 3949599, 2002-03-22.
14. H. Shibata, S. Naka, T. Ueda, Semiconductor device and manufacturing method, JP Patent 4094324, 2002-04-05.
15. N. Makita, M. Nakazawa, H. Ohnuma, T. Matsuo, Semiconductor device and method of manufacturing the same, U. S. Patent 7625786, 2001-06-28.
16. K. Kato, A. Isobe, H. Miyairi, S. Yamazaki, Semiconductor memory cell and semiconductor memory device, U. S. Patent 6812491, 2002-03-22.
17. H. Miyagawa, H. Kusai, R. Takaishi, T. Kawai, Y. Kamimuta, T. Murakami, K. Ariyoshi, T. Asano, M. Goto, M. Fujiwara, Y. Mitani, T. Obu and H. Aochi, "Metal-Assisted Solid-Phase Crystallization Process for Vertical Monocrystalline Si Channel in 3D Flash Memory," *IEEE IEDM,* pp. (IEDM19-650)-(IEDM19-653), 2019.
18. F. Mo, Y. Tagawa, C. Jin, M. Ahn, T. Saraya, T. Hiramoto, and M. Kobayashi, "Experimental Demonstration of Ferroelectric HfO2 FET with Ultrathin-body IGZO for High-Density and Low-Power Memory Application," *IEEE VLSI*, pp. T42-43, 2019. 10.23919/VLSIT.2019.8776553
19. M. Cho, H. Seol, A. Song, S. Choi, Y. Song, P. S. Yun, K. Chung, J. U. Bae, K. Park, and J. K. Jeong, "Comparative Study on Performance of IGZO Transistors With Sputtered and Atomic Layer Deposited Channel Layer," *IEEE T-ED*, Vol. 66, No. 4, 2019. 10.1109/TED.2019.2899586
20. T. Onuki, W. Uesugi, A. Isobe, Y. Ando, S. Okamoto, K. Kato, T. R. Yew, J. Y. Wu, C. C. Shuai, S. H. Wu, J. Myers, K. Doppler, M. Fujita, and S. Yamazaki, "Embedded Memory and ARM Cortex-M0 Core Using 60-nm C-Axis Aligned Crystalline Indium–Gallium–Zinc Oxide FET Integrated With 65-nm Si CMOS," *IEEE JSSC*, Vol. 52, No. 4, 2017. 10.1109/JSSC.2016.2632303
21. N. Kimizuka and S. Yamazaki, "Physics and Technology of Crystalline Oxide Semiconductor CAAC-IGZO Fundamentals," *WILEY*, 2017.
22. T. Fung, C. Chuang, C. Chen, K. Abe, R. Cottle, M. Townsend, H. Kumomi, and J. Kanicki, "Two-dimensional numerical simulation of radio frequency sputter amorphous In–Ga–Zn–O thin-film transistors" *J. Appl. Phys.*, Vol. 106, No. 084511, 2009. https://doi.org/10.1063/1.3234400
23. I. M. Choi, M. J. Kim, N. On, A. Song, K. Chung , H. Jeong, J. K. Park, and J. K. Jeong, "Achieving High Mobility and Excellent Stability in Amorphous In–Ga–

Zn–Sn–O Thin-Film Transistors," *IEEE T-ED*, Vol. 67, No.3, 2020. 10.1109/TED.2020.2968592

24. S. Samanta, U. Chand, S. Xu, K. Han, Y. Wu, C. Wang, A. Kumar, H. Velluri, Y. Li, X. Fong, A. V. Thean, and X. Gong, "Low Subthreshold Swing and High Mobility Amorphous-Indium-Gallium-Zinc-Oxide Thin-Film Transistor with Thin HfO2 Gate Dielectric and Excellent Uniformity," *IEEE EDL* (Early Access), 2020. 10.1109/LED.2020.2985787

25. J. H. Song, N. Oh, B. D. Anh, H. D. Kim, and J. K. Jeong, "Dynamics of Threshold Voltage Instability in IGZO TFTs: Impact of High Pressurized Oxygen Treatment on the Activation Energy Barrier," *IEEE T-ED*, Vol. 63, No. 3, 2016. 10.1109/TED.2015.2511883

26. X. D. Huang, J. Q. Song, and P. T. Lai, "Positive Gate Bias and Temperature-Induced Instability of α-InGaZnO Thin-Film Transistor With ZrLaO Gate Dielectric," *IEEE T-ED*, Vol. 63, No. 5, 2016. 10.1109/TED.2016.2541319

27. S. Choi , S. Park, J. Kim, Y. Seo, H. J. Shin, Y. S. Jeong, J. U. Bae , C. H. Oh, D. M. Kim, S. Choi, and D. H. Kim, "Positive Bias Stress Instability of InGaZnO TFTs With Self-Aligned Top-Gate Structure in the Threshold-Voltage Compensated Pixel," *IEEE EDL*, Vol. 41, No. 1, 2020. 10.1109/LED.2019.2954543

28. H. Lue, T. Yeh, P. Du, R. Lo, W. Chen, T. Hsu, C. Huang, G. Lee, C. Chen, Y. Jiang, M. Hung, Y. Su, L. Liang, C. Hu, C. Chiu, K. Wang, and C. Lu, "A Novel Double-Density Hemi-Cylindrical (HC) Structure to Produce More than Double Memory Density Enhancement for 3D NAND Flash," *IEEE IEDM*, pp. 28.2.1-28.2.4, 2019. 10.1109/IEDM19573.2019.8993461

68

Highly stable self-aligned coplanar indium gallium zinc oxide thin-film transistors for high-resolution OLED TV and mobile displays

J. B. Kim[a], Y. C. Tsai[a], R. Lim[a], Z. Wang[a], M. Hao[a], S. W. Wang[b], J. W. Park[a], L. Zhao[a], M. Bender[c], D. K. Yim[a], and S. Y. Choi[a]

[a] AKT Display, Applied Materials, 3101 Scott Blvd, Santa Clara, California, 95054 USA
[b] Applied Materials Taiwan, No. 12, Nanke 7th Road, Tainan Science-Based Industrial Park, Taiwan, R.O.C.
[c] Applied Materials GmbH & Co. KG, Siemensstr. 100, 63755 Alzenau, Germany

> Highly stable self-aligned coplanar top-gate InGaZnO (IGZO) thin-film transistors (TFTs) with excellent threshold voltage (Vth) uniformity are developed by optimizing gate insulator film properties and controlling oxygen/hydrogen diffusion into IGZO channel. The TFTs do not show notable negative Vth shift and Vth non-uniformity from channel length 10 μm down to 3 μm. The TFTs also show no notable Vth shift during a humidity test in a chamber at 85 °C/85 % after deposition of passivation layers or even after 1-μm-thick thin-film encapsulation (TFE) SiNx layer on the top of the passivation layer due to the high quality/low hydrogen 2nd SiNx passivation layer deposited on the top of 1st SiOx passivation layer.

Introduction

Indium gallium zinc oxide (IGZO) thin-film transistors (TFTs) have attracted much attention because of their high electron mobility values (> 10 cm^2/Vs) and because they can be processed at low temperature to produce large-area displays with the potential of low production costs [1]. Bottom-gate (BG) TFTs using back channel etch (BCE) and etch stopper (ES) types commercially used for mid-to-high resolution liquid crystal displays (LCDs) limit high-speed driving in high-resolution OLED displays due to larger parasitic capacitance from overlap area between gate and source/drain electrodes in TFTs. Therefore, it's necessary to use self-aligned top-gate (TG) oxide TFT structures with smaller channel length for high-resolution, high-speed, low-power, and narrow-bezel displays. The oxide TFTs need good electrical stability under positive/negative bias temperature stress (P/NBTS) with decent TFT properties for peripheral gate driver integrated large size organic light emitting diode (OLED) TVs. In general, TG oxide TFT needs to show decent initial threshold voltage (Vth) close to 0 V and Vth uniformity variation of < 1 V with decent sub-threshold slope (SS) value of < 0.5 V/decade, field effect mobility (μ_{FE}) around 10 cm^2/Vs, and low off-leakage current (I_{OFF}) of $< 10^{-12}$ A for proper panel circuit operation. TG oxide TFTs are currently used as pixel and gate driver in panel (GIP) TFTs in commercial AMOLED displays such as TVs. Reliability of the GIP buffer TFTs under positive bias temperature stress (PBTS) becomes more important because AMOLED panel reliability is determined by the reliability of GIP TFTs, especially for pull-up and pull-down buffer TFTs. AMOLED displays employing GIP TFTs also needs to use shorter channel length of TFTs to reduce device size for narrow display bezel and to increase drain-to-source current (I_{DS}) for low voltage and low power operations. J. Y. Noh, et al demonstrated 5.5 mm-thick narrow bezel 55" 4K UHD AMOLED TVs employing GIP circuits with 4.5 μm channel length IGZO TFTs by optimizing ΔL (the oxygen vacancies diffusion length from n+ IGZO region to channel region) [2]. However, further reduction of channel length less than 4.5 μm in GIP TFTs is preferred but is not easy due to larger ΔL and Vth negative shift with larger Vth non-uniformity. It is well known that negative Vth shift starts by decreasing channel length when channel length becomes very small. When channel length becomes smaller than 10 μm, short channel effect affects to TFT uniformity as cause of the negative Vth shift due to drain bias induced barrier lowering (DIBL) and/or unintentional doping into IGZO near n+ doped source-drain electrodes [2, 3]. The channel length should be as small as possible for narrow bezel, low-power, high-speed and high-resolution displays. Therefore, minimizing short channel effect by decreasing ΔL during TFT fabrication is necessary. On the other hand, current oxide TFTs are facing challenges in the use of mobile display applications such as smart phone due to its low mobility and worse electrical stability with difficulty of Vth control especially with shorter channel TFT compared to low temperature poly-Si (LTPS) TFT. However, Apple Inc. recently applied a new TFT integration technology for the display backplane pixel circuit of mobile iWatch 4 by using TG oxide TFT as a switching device with TG LTPS TFTs as switching and driving TFTs, which allows various frame frequency operations with variable refresh rate (VRR) function even with the frame frequency of 1Hz without flickering display image due to low off leakage current of oxide TFT for power saving. The new TFT integration technology is called as LTPO (LTPS + Oxide) by integrating both oxide TFT and LTPS TFT on the same substrate [4]. TG oxide TFTs for display applications such as OLED TVs and LTPO mobile displays currently use longer channel length (> 3 μm) due to negative Vth shift and Vth non-uniformity issues with smaller channel length, which is mainly caused by hydrogen diffusion near source/drain regimes during device integration.

In this work, we fabricated TG IGZO TFTs by using Applied Material's CVD/PVD tools to achieve excellent electrical stability and Vth uniformity by balancing oxygen and hydrogen diffusion into oxide channel layer such as IGZO. Suppressing carrier densities at the interface between gate insulator (GI) and IGZO by increasing oxygen diffusion with IGZO pre-annealing in air and TFT post anneal in air, and by minimizing vacuum annealing impact during GI deposition makes the IGZO TFTs more resistive with more positive Vth shift. On the other hands, more hydrogen diffusion into IGZO from GI, ILD, and passivation layers during device integration or TFT post annealing makes the IGZO TFTs more conductive with more negative Vth shift and worse Vth uniformity, but it helps to improve electrical stability with smaller Vth shift under PBTS by passivating defects inside IGZO. Therefore, there is trade-off between Vth uniformity and electrical stability due to the dominance of the diffusion between oxygen and hydrogen. By balancing the diffusion of oxygen and hydrogen, the IGZO TFTs can minimize negative Vth shift even with shorter channel for high-resolution display applications and achieve good electrical stability for pixel and peripheral gate driver circuits integrated in display panel.

Device Fabrication

Top-gate self-aligned coplanar IGZO TFTs were fabricated as shown in Fig 1. All the dielectric layers (buffer/gate insulator/interlayer dielectric layers) and IGZO channel layer were optimized and deposited by AKT PECVD and PVD systems, respectively. Cr was used for both gate and source-drain electrodes and it was deposited through AKT PVD system. First, a 350-nm-thick SiOx as a buffer layer was deposited on Si wafer. A 30-nm-thick IGZO layer as a channel layer was then deposited and annealed in air. The IGZO layer was patterned with diluted 1% HCl by wet etch. After this, a 200-nm-thick SiOx as a gate insulator layer was deposited on the top of IGZO layer followed by the deposition of a 170-nm-thick Cr gate electrode. After patterning gate electrode by dry etch, gate insulator was patterned using gate electrode as a mask and then IGZO n+ regime is formed by Helium (He) plasma treatment. After that, a 400-nm-thick interlayer dielectric (ILD) layer was deposited and contact was opened by dry etch. Then, a 230-nm-thick Cr as source-drain electrodes was deposited. After patterning source/drain electrodes by dry etch, devices were annealed in air. After annealing in air, electrical properties and stabilities of the oxide TFTs are validated. For the further validation of the oxide TFTs as water (H2O) or gas (oxygen, hydrogen) barriers, single SiOx and dual SiOx/SiNx passivation (PAS) layers are deposited on the top of source/drain electrodes. After that, devices were annealing in air.

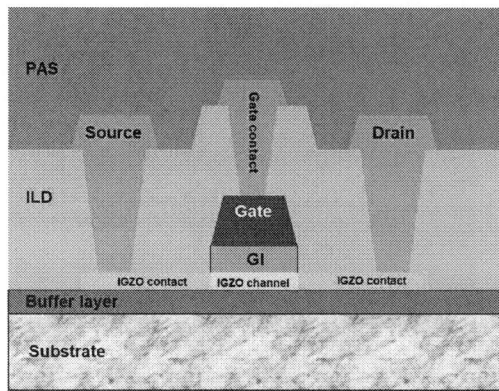

Figure 1. Device structure of top-gate self-aligned coplanar IGZO TFT

Results and Discussion

Fig. 2 is illustrative of the new manufacturing tools being developed to support oxide TFT technology. This semi-dynamic cluster PVD tool is targeted for Gen 4.5 (Gen 6) applications, suitable for R&D with small footprint, scaleable for Gen6 towards mobile applications. Semi-dynamic platform provides mura-free oxide films, based on proven rotary PVD architecture, easy to control plasma ignition and ion bombardment for high quality films as well as improved defectivity.

Figure 2. A new Gen 4.5 semi-dynamic cluster PVD tool for oxide TFT manufacturing. This system features a rotary array deposition source and horizontal substrate movement for improved uniformity.

For the fabrication of TG TFTs, various types of recipes are used with different deposition temperature and different radio frequency (RF) power during GI deposition. Fig. 3 shows film properties of GI with different temperature (225 °C, 270 °C, 300 °C) and different radio frequency (RF) power (800 W, 1600 W, 2400 W). To verify film properties, SiOx peak position value and wet etch rate are checked by FTIR and buffer oxide etch solution, respectively. Better film quality films typically show the values of higher SiOx peak position and lower SiOx wet etch rate. SiOx GI films deposited at high temperature and/or high RF power show better film quality with the values of higher SiOx peak position and lower wet etch rate than SiOx GI film at low temperature and/or low RF power.

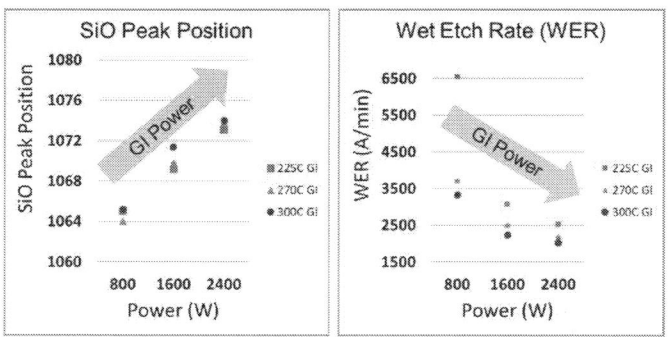

Figure 3. SiO peak position and wet etch rate (WER) of gate insulator (GI) deposited by using AKT Gen 4.5 PECVD system with different temperature and different RF power

TG TFTs are fabricated by varying GI deposition temperature from 225 °C to 290 °C and RF power from 1200 W to 2400 W. To investigate GI quality impacts for the properties and stabilities of oxide TFTs, five types of GI films such as GI type A, B, C, D, and E are deposited. Here, GI type A, B, and C are deposited at different temperature 225 °C, 270 °C, and 290 °C, respectively, at a fixed RF power of 2400W and GI type B, D, and E are deposited at different RF power 2400W, 1600W, and 1200W, respectively, at a fixed temperature of 270 °C. Fig. 4 shows transfer characteristics from the TFTs with the five types of GI films. After small tuning of GI recipes considering carrier densities between GI and IGZO, electrical properties are optimized. The electrical characteristics of the TFTs were measured at room temperature using an Agilent 4156C semiconductor analyzer. The field effect mobility, μ_{FE} (cm²/Vs), was extracted from transconductance ($\partial I_{DS}/ \partial V_{GS}$) at $V_{GS} = Vth + 10$ V, where I_{DS} is the drain-to-source current and V_{GS} is gate-to-source voltage. The threshold voltage, Vth (V), is determined from $I_{DS} = 10^{-9} \times$ (W/L) (A) at $V_{GS} = Vth + 10$ V. The subthreshold voltage swing, SS (V/decade), is extracted from the minimum value of [$\partial Log(I_{DS})/ \partial (V_{GS})]^{-1}$ at $V_{DS} = +1$ V. The leakage current, I_{OFF} (A), is determined from I_{DS} at $V_{GS} = -7$ V when $V_{DS} = +10$ V. The TG oxide TFTs from the five different types of GI films show good electrical properties such as field effect mobility (μ_{FE}) of > 10 cm²/Vs, sub-threshold slope (SS) of < 0.35 V/decade, and threshold voltage (Vth) close to 0 V with low off leakage current (I_{OFF}) around 1x10⁻¹² A or less. For further investigation of the electrical performance in TFTs, bias temperature stress (BTS) was provided to the TFTs. For positive bias temperature stress (PBTS) and negative bias temperature stress

(NBTS), positive bias voltage of +30 V and negative bias voltage of -30 V were applied to gate electrode, respectively, by connecting source and drain electrodes to ground at the substrate temperature of 60 °C. During PBTS and NBTS, Vth values were monitored every 600 seconds by sweeping gate-to-source voltage (V_{GS}) from -20 V to +20 V at drain-to-source voltage (V_{DS}) of +1 V up to 1 hour.

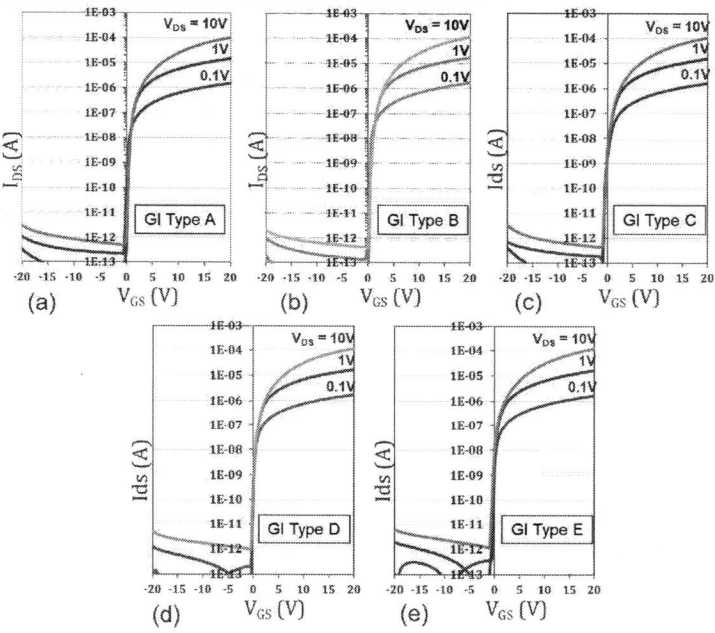

Figure 4. Transfer characteristics of TG oxide TFTs with channel width of 40 μm and channel length of 10 μm from different (a) GI type A, (b) GI type B, and (c) GI type C deposited at different temperature 225 °C, 270 °C, 290 °C with a fixed RF power of 2400 W and from different (d) GI type D and (e) GI type E deposited at different RF power 1600 W, 1200 W with a fixed temperature of 270 °C.

As shown in Fig. 5(a), PBTS from GI type B (0.66 V) is better than PBTS from GI type A (2.29 V). Further improvement of PBTS is observed from GI type C (0.16 V). On the other hand, PBTS from GI type D (0.54 V) is better than PBTS from GI type B (0.66 V) as shown in Fig. 5(b). Further improvement of PBTS is observed from GI type D (0.40 V). Improvement of PBTS is mainly caused by more hydrogen partial diffusion at higher temperature and/or lower RF power during GI deposition. Decent NBTS values are achieved from TFTs with different GI type A (0.26 V), GI type B (0.04 V), GI type C (0.06 V), GI type D (0.14 V), and GI type E (0.17 V). During PBTS and NBTS, TFT transfer characteristics show parallel shift of Vth without degrading mobility and SS values. Therefore, the parallel shift of Vth is mainly caused by charge trapping/de-trapping at the interface between GI and IGZO films under bias stress. Better electrical stability from high temperature GI than low temperature GI is indicating less charge trapping/de-trapping into

GI film under bias stress. Table 1. shows average electrical properties and stabilities from TFTs with different GI type A, B, C, D, and E.

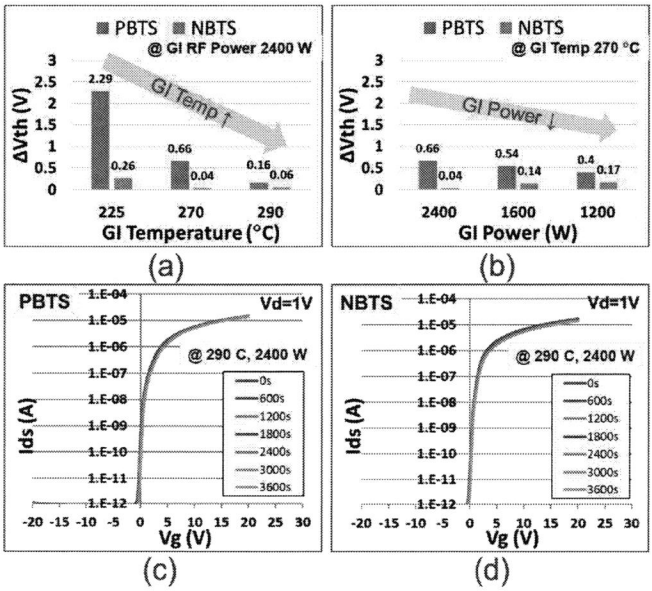

(a)   (b)

(c)   (d)

Figure 5. Threshold voltage shift under PBTS and NBTS from oxide TFTs (a) with different GI temperature 225 °C, 270 °C, 290 °C at a fixed RF power 2400 W and (b) with different RF power 2400 W, 1600 W, 1200 W at a fixed GI temperature 270 °C. Transfer characteristics were monitored every 600 seconds up to 1 hour by sweeping gate-to-source voltage at drain-to-source voltage of +1 V under (c) PBTS and (d) NBTS from TFTs from GI type C with temperature of 290 °C and RF power of 2400 W.

TABLE 1. Average electrical properties and stabilities from TFTs with GI type A, B, C, D, and E.

Items	Unit	Type A	Type B	Type C	Type D	Type E
GI Temp.	°C	225	270	290	270	270
GI Power	W	2400	2400	2400	1600	1200
Vth	V	0.30	0.35	0.02	0.10	0.06
SS	V/decade	0.18	0.19	0.17	0.16	0.17
μ_{FE}	cm²/Vs	11.4	12.2	11.5	13.2	12.0
I_{OFF}	pA	0.68	0.84	0.56	1.5	1.8
PBTS	V	2.29	0.66	0.16	0.54	0.4
NBTS	V	0.26	0.04	0.06	0.14	0.17

In order to investigate the channel length dependency of Vth, Vth values were extracted from transfer characteristics at V_{DS} = +1 V with various TFTs by fixing channel width of 12 μm and varying the channel length from 10 μm down to 3 μm. Transfer characteristics from TFTs with GI type A, B, and D are shown in Fig. 6(a), 6(b), and 6(c), respectively. Transfer characteristics of TFTs with the channel length from 10 μm down to 3 μm do not show notable negative Vth shift, which indicates that unintentional doping effect from n+ doped source and drain regimes into IGZO channel are not dominant even with channel length of 3 μm. The differences between maximum Vth and minimum Vth, Vth (Max–Min), from GI type A, B, and D are 0.35 V, 0.48 V, and 1.20 V, respectively. Slightly larger Vth variation and non-uniformity are observed from higher temperature and/or lower RF power due to more hydrogen partial diffusion during GI deposition than lower temperature and/or higher RF power. From previous reports, larger amount of oxygen vacancies at the interface between GI and IGZO or in IGZO are significantly changing threshold voltages as channel length become less than 10 μm, which make IGZO channel in TFTs more conductive as a cause of negative Vth shift [2, 3]. The impact of oxygen vacancies become stronger as channel length become smaller. Therefore, the ratio (L_E/L) between effective channel length (L_E) and mask channel length (L) becomes smaller by decreasing channel length. Interestingly, our TFTs do not show notable Vth shift with channel length from 10 μm down to 3 μm. To investigate short channel length effect, the difference (2ΔL) between mask channel length (L) and effective channel length (L_E) was extracted by the transmission line method (TLM) [5, 6].

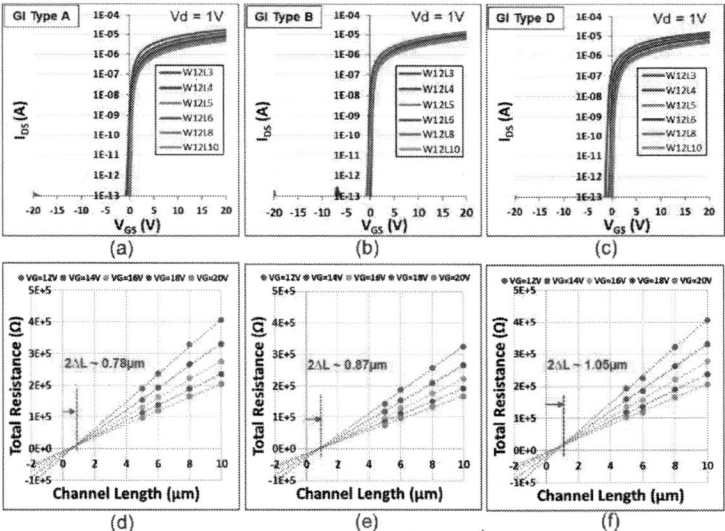

Figure 6. Transfer characteristics (a, b, c) of TG TFTs with L = 10 μm down to 3 μm at fixed width of 12 μm, and total TFT resistances (d, e, f) extracted by transmission line method with channel length down from 10 μm to 5 μm with GI type A, B, and D.

Total resistance in TFTs was extracted from the measured I_{DS} values by varying V_{GS} from 12 V to 20 V with channel length from 10 μm down to 5 μm as shown in Fig. 6(d), 6(e), and 6(f). Extracted 2ΔL values from GI type A, B, and D are 0.78 μm, 0.87 μm, and 1.05 μm, respectively. This result indicates that short channel length effect from our TFTs is not critical issues and the oxygen vacancies near n+ source-drain regimes are well controlled during fabrication processes. Slightly larger channel length reduction from GI type B and D than type A is caused by more hydrogen partial diffusion from GI deposited at high temperature and/or low RF power. Negative Vth shift from TFTs can be controlled by decreasing carrier density in the regimes of both channel and n+ source and drain contacts by TFT post annealing in air. However, decreasing carrier density increases both contact resistance and channel resistivity, which reduces mobility by decreasing I_{DS} in TFT. Our TFTs show low contact resistance (2Rc) values of less than 20 kΩ even after final TFT anneal in air, which allows TFTs to maintain high mobility and high I_{DS} without current crowding. During TFT fabrication, we first tuned Vth position values by carefully controlling carrier density at the interface between GI and IGZO and/or inside IGZO channel area and then significantly reduced IGZO resistivity near source-drain contact regime by He plasma treatment after GI etch. Further tuning of Vth was performed with TFT post annealing in air. We observed significant changes of initial Vth position according to various GI deposition conditions. If there are too much carriers at the interface between GI and IGZO and/or inside IGZO channel after GI deposition, TFT post annealing in air with strong oxygen diffusion into IGZO channel can reduce the carrier density to control Vth, but it is not easy to control Vth without sacrificing mobility and I_{DS}. Therefore, carrier density can be controlled by IGZO post annealing in air before GI deposition, GI recipe tuning during GI deposition, and TFT post annealing in air after TFT fabrication. Note that hydrogen can diffuse into IGZO from both GI and ILD layers during deposition and TFT post annealing, which causes negative Vth shift with larger Vth non-uniformity. Understanding of how IGZO TFT is affected by hydrogen and oxygen diffusion will be meaningful for the control of both Vth non-uniformity and device stability.

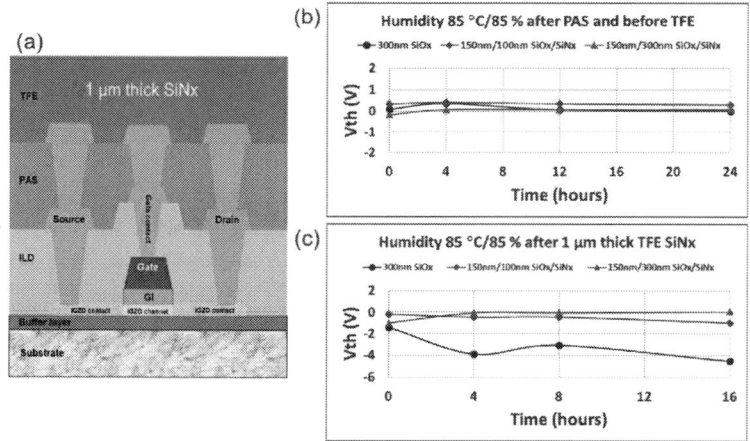

Figure 7(a). Device structure of self-aligned coplanar oxide TFTs for environmental stability validation by adding passivation and TFE SiNx layers on the top of source/drain electrodes. 7(b). Environmental stability of the oxide TFTs in a humidity chamber at 85 °C/ 85% after deposition of single and dual passivation layers. 7(c). Environmental stability of the oxide TFTs in a humidity chamber at 85 °C/ 85 % after deposition of 1 µm thick TFE SiNx layer on the top of passivation layers.

Figure 7(a) shows a device structure of TG oxide TFTs for environmental stability validation by adding passivation (PAS) layer on the top of source/drain electrodes and then 1-µm-thick SiNx thin-film encapsulation (TFE) layer on the top of the PAS layers. To validate the performance of the PAS layers as water (H_2O) or gas barrier from oxygen or hydrogen, single SiOx and dual SiO/SiNx passivation layers are deposited on the top of source/drain electrodes. Here, a single 300-nm-thick SiOx PAS layer is deposited on the top of source/drain electrodes. For the dual PAS layers, SiOx/SiNx layers are deposited on the top of source/drain electrodes by depositing two different thicknesses (100 nm, 300 nm) of SiNx layers on the top of a 150-nm-thick SiOx layer. After deposition of the PAS layers, TFTs were annealing in air. Electrical properties such as μ_{FE}, SS, Vth, and I_{OFF} from the TFTs do not show notable changes of μ_{FE}, SS, Vth, and I_{OFF} before and after deposition of PAS layers with small negative Vth shift after deposition of the PAS layers. Then, the TFTs were exposed in a humidity chamber at 85 %/85 °C up to 24 hours to monitor electrical properties such as Vth. The TFTs from both single and dual PAS layer also do not show notable Vth shift as shown in Fig. 7(b). After 24 hours in a humidity chamber, 1-μm-thick SiNx layer as a TFE layer is deposited at a low temperature of 80 °C to monitor hydrogen diffusion impact into IGZO. The TFE SiNx layer deposited at low temperature contains high hydrogen percentage of > 40%. After deposition of the TFE SiNx layer, TFTs with single 300-nm-thick SiOx passivation layers show Vth of -1.42 V with negative shift of around 1.4 V. However, TFTs with dual SiOx/SiNx passivation layers show less negative Vth shift by blocking hydrogen diffusion into IGZO from the TFE SiNx layer due to the 2nd SiNx passivation layer. Here, we used a high quality/low hydrogen SiNx film as the 2nd PAS layer. The 2nd PAS layer works as both a source and a blocker of hydrogen. Less negative shift from the dual SiOx/SiNx passivation layer means that the 2nd SiNx PAS layer works as an excellent blocker of hydrogen from the top layer TFE SiNx due to its high quality with less diffusion of hydrogen into IGZO through the 1st SiOx PAS, ILD and GI layers due to its low amount of hydrogen. TFTs with 150-nm/100-nm-thick and 150-nm/300-nm thick dual SiOx/SiNx PAS layers show smaller negative Vth shift values of around 0.6 V and 0.97 V, respectively, than TFTs with 300-nm-thick single SiOx PAS layer after deposition of the TFE SiNx layer. More negative shift from thicker 300-nm-thick SiNx layer than thinner 100-nm-thick SiNx layer might be due to higher amount of hydrogen in the thicker SiNx layer. For the further test for environmental stability, TFTs are exposed in the humidity chamber at 85 %/85 °C up to 16 hours. As expected, TFTs with single SiOx PAS layer show large negative Vth shift of 3.09 V from -1.42 V to -4.51 V. However, TFTs with dual 150-nm/100-nm-thick SiOx/SiNx PAS layer show less negative Vth shift of 0.76 V from -0.19 V to -0.95 V. Interestingly, TFTs with dual 150-nm/300-nm-thick SiOx/SiNx PAS layer show positive Vth shift of 0.98 V from − 0.90 V to +0.08 V because thicker SiNx passivation layer might be working as better blocker of hydrogen from the TFE SiNx layer. Table 2 shows the summary of Vth changes in oxide TFTs after deposition of PAS layer, after 24 hours in a humidity chamber at 85 °C/ 85% before deposition of TFE SiNx layer, after deposition of 1-μm-thick TFE SiNx layer, and after 16 hours in a humidity chamber at 85 °C/ 85%.

TABLE 2. Threshold voltage (Vth) values of the oxide TFTs after deposition of passivation (PAS) layer, after 24 hours in a humidity chamber at 85 °C/ 85% before deposition of TFE SiNx layer, after deposition of 1 μm thick TFE SiNx layer, and after 16 hours in a humidity chamber at 85 °C/ 85%.

PAS layers	300 nm SiOx	150 nm/100 nm SiOx/SiNx	150 nm/300 nm SiOx/SiNx
Vth after deposition of PAS layer	0.09 V	0.34 V	-0.17 V
Vth after 24 hrs in humidity chamber	0.01 V	0.41 V	0.07 V
Vth after deposition of TFE SiNx layer	-1.42 V	-0.19 V	-0.90 V
Vth after 16 hrs in humidity chamber	-4.51 V	-0.95 V	+0.08 V

Conclusion

Highly stable top-gate self-aligned coplanar IGZO TFTs with decent electrical properties are demonstrated by optimizing gate insulator film properties. The TFTs do not show notable short channel effect and channel length dependency even with 3 μm channel length. It is achieved by controlling hydrogen and oxygen diffusion into IGZO channel near source-drain electrodes during device fabrication. Although TFTs with 3 μm channel length are enough for most LCD and AMOLED display applications, shorter channel length less than 3 μm is desirable to realize low-power, high-speed, and high-resolution displays. The TFTs also shows excellent environmental stability without showing notable Vth shift under humidity test in a chamber at 85 C/85 %.

References

1. K. Nomura, H. Ohta, A. Takagi, T. Kamiya, M. Hirano, and H. Hosono, *Nature*, **432**, 488-492 (2004).

2. J. H. Noh, D. M. Han, W. C. Jeong, J. W. Kim, and S. Y. Cha, *SID Symposium Digest of Technical Papers*, **21/1**, 288-290 (2017).

3. D. H. Kang, J. U. Han, M. Mativenga, S. H. Ha, and J. Jang, *Applied Physics Letter*, **102**, 053508 (2013).

4. T. K. Chang, C. W. Lin, and S. Chang, *SID Symposium Digest of Technical Papers*, **51/1**, 545-548 (2019).

5. J. B. Kim, R. Lim, J. Wang, L. Zhao, Y. Tsai, M. Bender, D. K. Yim, and S. Y. Choi, *SID Symposium Digest of Technical Papers*, **62/1**, 874-877 (2019).

6. S. Park, E. N. Cho, and I. Yun, *Solid-State Electronics*, **75**, 93-96 (2012).

Chapter 3

H03 – Processes

80

Interpretation of Donor Activation in Boron and Argon Implanted Self-Aligned Bottom-Gate IGZO TFTs

M. S. Kabir[a], R. R. Chowdhury[b], R. G. Manley[c], and K. D. Hirschman[a,b]

[a] Department of Microsystems Engineering
[b] Department of Electrical and Microelectronic Engineering,
Rochester Institute of Technology, Rochester, New York, 14623, USA
[c] Corning Incorporated, Science & Technology Research,
Corning, New York, 14831, USA

This work provides an interpretation of donor activation in self-aligned bottom-gate (SA-BG) Indium-Gallium-Zinc Oxide (IGZO) TFTs with ion-implantation of boron ($^{11}B^+$) and argon ($^{40}Ar^+$) species as the source/drain treatment. Device fabrication strategies that switched the order of ion implantation and passivation annealing processes were investigated. Hypotheses in the mechanisms involved with donor activation have been developed from observed similarities and differences in the electrical operation of SA-BG TFTs fabricated using both implant-last and anneal-last approaches. Results suggest a defect-induced mechanism is responsible for donor behavior associated with argon implantation, whereas donor behavior associated with boron is attributed to the formation of an electrically active donor species involving chemical bonding. A detailed discussion on developed arguments arrived at through analysis of TFT electrical characteristics is presented. Materials analysis in support of the interpretation and associated arguments is in progress.

Introduction

The source/drain electrodes in IGZO TFTs (thin-film transistors) can be direct metal overlapping the IGZO region. However, to achieve ohmic behavior with minimal resistance and ensure tolerance to overlay error, several microns of gate overlap is required. IGZO TFTs with self-aligned channel regions have advantages in reduced parasitic capacitance and stage delay, and a reduction in overhead real estate. Several techniques have been used to selectively form conductive IGZO regions including the introduction of hydrogen (1, 2), plasma exposure (3) and ion implantation (4). We have recently presented an original investigation on the application of boron ion implantation for a self-aligned top-gate (SA-TG) IGZO TFT (5). As an extension, the following work provides an interpretation of donor activation induced by ion-implantation of boron ($^{11}B^+$) and argon ($^{40}Ar^+$) species as the source/drain treatment applied to SA IGZO devices.

The bottom-gate staggered electrode configuration was used for this investigation, utilizing the opaque gate electrode as a mask for a through-wafer exposure of positive photoresist. This technique provides a perfectly aligned implant mask and averts charge accumulation on a floating metal top-gate during the ion implant process, which has been suspected of subjecting the gate dielectric to a high electric field and the creation of ionized defects (5). Utilizing an "implant-last" strategy, both boron and argon implanted SA bottom-gate (SA-BG) devices exhibited electrical characteristics consistent with non-SA devices, with minimal impact of added series resistance. However, there were noted differences in the device operation that became pronounced following thermal stability testing below 200 °C. SA-BG TFTs that utilized an "anneal-last" strategy resulted in a much lower degree of active donors, with a significant compromise in electrical characteristics. Hypotheses in the mechanisms involved with donor activation have been developed from observed similarities and differences in the electrical operation of SA-BG TFTs fabricated using both implant-last and anneal-last strategies. Experimental details, electrical characteristics, and an interpretation of donor activation behavior are presented.

Experimental Details

Using a glass substrate, a 50 nm thick Mo gate electrode was sputtered and patterned, followed by a PECVD SiO_2 gate dielectric, which was densified for 2 hours at 600 °C in nitrogen. A 50 nm IGZO layer was sputtered using an $InGaZnO_4$ (1:1:1:4) target in an argon ambient with 7% oxygen, and then patterned and etched using dilute HCl. The S/D contact metal (100 nm Mo/Al bilayer) was sputtered and patterned using a lift-off process. A 50 nm PECVD SiO_2 passivation oxide was then deposited, followed by an O_2 passivation anneal at 400 °C and an ALD Al_2O_3 capping layer. The wafer was then coated with AZ MIR 701 positive resist following an HMDS vapor prime at 140 °C. Back-side illumination with broadband spectrum was done on a Suss MA150 contact aligner, with black felt used as an underlying absorption layer to avoid reflections. After exposure, the resist was developed for 45 sec in Microposit MF CD-26 developer solution and hotplate baked for 60 s at 140 °C. The samples were then ion implanted for the S/D activation treatment following the implant-last strategy. Finally, the resist was removed, and the S/D contact regions were patterned and opened using 10:1 buffered HF. For the anneal-last strategy, the S/D implant was done before the O_2 passivation anneal. Process variants included the implant dose (i.e. ϕ and 2ϕ fluence) introduced in the implant-last approach.

Electrical testing was done using a B1500 semiconductor parameter analyzer. TFT channel dimensions were of width W = 24 µm and length as indicated. All I_D-V_{GS} transfer characteristics presented were taken with a gate voltage up-sweep and medium measurement integration time unless otherwise noted, with low-drain and high-drain bias conditions at 0.1 V and 10 V, respectively. Thermal stability testing was done using a hotplate temperature setting at 175 °C for 1-hour time intervals.

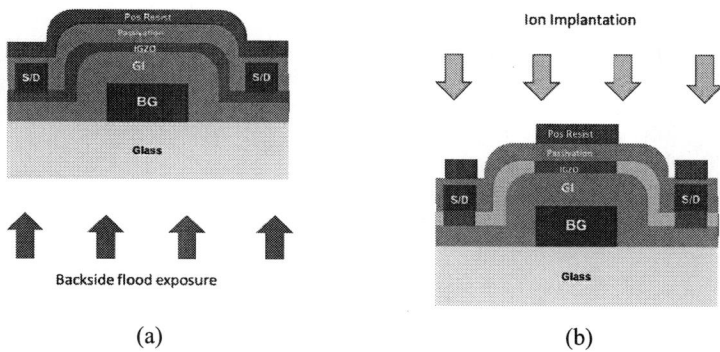

(a)   (b)

Figure 1. Cross-sectional schematic of an SA-BG device during (a) back-side flood exposure and (b) S/D implant treatment using the self-aligned resist polygon as an implant mask, where the orange gap region represents the implanted IGZO, which serves as the working source/drain electrode. Note that in this process scheme the implant is blocked in both channel and S/D metal contact regions.

Results and Discussion

Argon Implanted SA-BG TFTs

Argon ions with energy of 80 keV were implanted with dose $\phi = 2 \times 10^{15}$ cm^{-2} with representative transfer characteristics shown in Figure 2. The initial implant-last characteristics shown in Figure 2a exhibited standard behavior, with the exception of a slight crossover effect attributed to localized trap states. After a 175 °C hotplate bake for 1 hour, the characteristics exhibited degradation with lower current drive and enhanced GIDL. The argon implanted SA-BG device with anneal-last strategy shown in Figure 2b demonstrated exceedingly poor performance in comparison to a non-SA-BG TFT fabricated on the same substrate.

Boron Implanted SA-BG TFTs

For implant-last devices, boron ions with energy of 35 keV were implanted at two different doses, $\phi = 4 \times 10^{15}$ cm^{-2} and $2\phi = 8 \times 10^{15}$ cm^{-2}, with transfer characteristics shown in Figure 3a and 3b, respectively. Figure 3a demonstrates a left shift in the characteristics for the shorter channel L = 6 μm device, compared to the L = 12 μm device. An analysis of the effective channel length using the Terada-Muta method (6) identified a process-induced offset $\Delta L \sim 3.5$ μm due to the molybdenum bottom-gate chemical etch. A proposed origin of the left shift in the $L_{eff} \sim 2.5$ μm device characteristics will be discussed further. When the boron dose was doubled to $2\phi = 8 \times 10^{15}$ cm^{-2}, the transfer characteristics degraded as shown in Figure 3b.

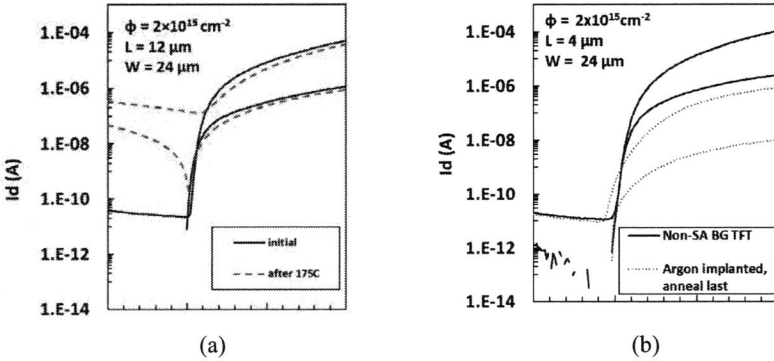

(a)   (b)

Figure 2: Comparison of $^{40}Ar^+$ implanted (E = 80 keV, $\phi = 2\times10^{15}$ cm^{-2}) SA-BG device transfer characteristics (a) implant-last strategy measured before (solid line) and after (dashed line) a 175 °C hotplate bake for 1 hour (b) characteristic overlay of an anneal-last device and a non-SA-BG TFT with the same channel dimensions as indicated.

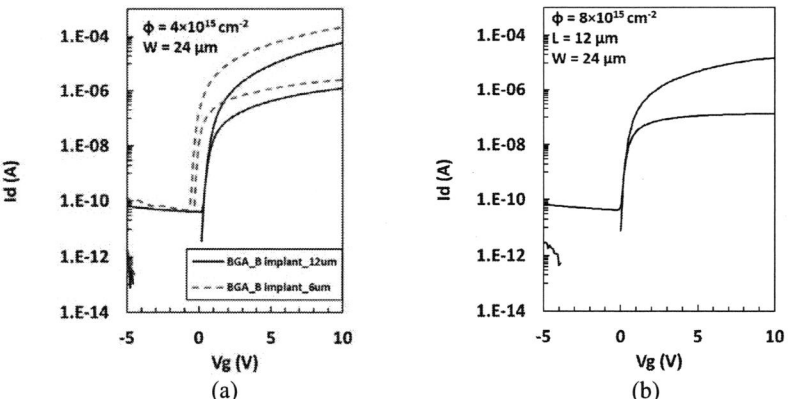

(a)   (b)

Figure 3: (a) Comparison of $^{11}B^+$ implanted SA-BG device characteristics with L = 6 μm (dashed line) and L = 12 μm (solid line) using the implant-last approach with $\phi = 4\times10^{15}$ cm^{-2} into 4 μm gap regions. (b) L = 12 μm device with a 2x dose increase ($2\phi = 8\times10^{15}$ cm^{-2}) demonstrating a degradation in charge injection that is pronounced at the lower drain bias.

After subjecting the devices to 175 °C hotplate bake for 1 hour, there was a slight shift in the transfer characteristics of the L = 12 μm device as shown in Figure 4a, which became significant as the cumulative time was increased to 2 hours and finally 4 hours. Figure 4b is an adjusted overlay of the original and 4-hour hotplate bake characteristics. Figure 4c shows the channel-length dependence of the amount of lateral left-shift induced by the 4-hour bake treatment.

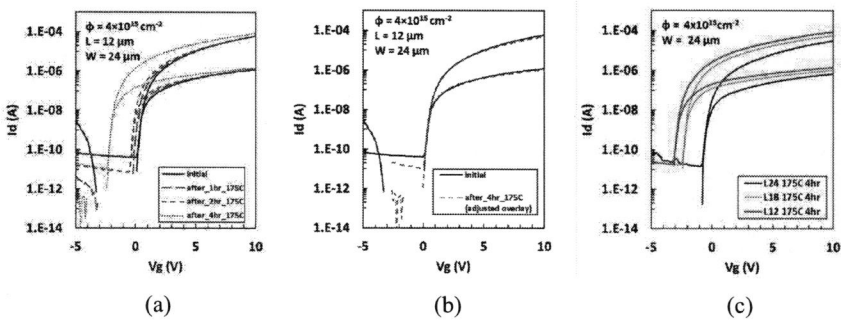

(a)   (b)   (c)

Figure 4: (a) Comparison of $^{11}B^+$ implanted SA-BG device characteristics with (a) L = 12 μm and $\phi = 4\times10^{15}$ cm^{-2} before (solid line) and after (dashed lines) successive 175 °C bakes, (b) adjusted overlay of the original and 4-hour hotplate bake characteristics. (c) channel-length dependence of shift induced by the 4-hour hotplate bake treatment.

Boron devices fabricated with an anneal-last strategy had exceedingly low current drive, with an overlay comparison of transfer characteristics from boron and argon anneal-last SA-BG TFTs along with a reference non-SA BG TFT shown in Figure 5.

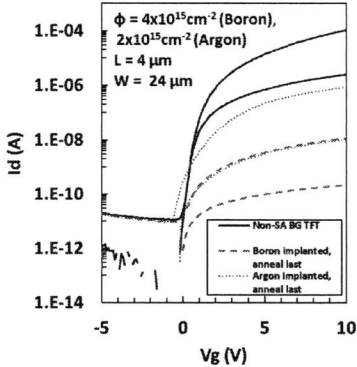

Figure 5: Replicate of Figure 2b, with the addition of the anneal-last boron implanted device characteristic for comparison to the other two devices shown previously.

Key Observations and Interpretation

The results of the argon implanted SA-BG TFTs support a defect-induced activation mechanism. Since argon is not chemically active, donor activation is presumably due to atom displacements, i.e. the creation of oxygen vacancies. The implant-last approach demonstrated a high level of donor activation, however the crossover characteristics shown in Figure 2a is attributed to localized defects in the transition between the active S/D regions and the channel. This was observed to be markedly worse on devices which received a higher argon implant dose (i.e. $2\phi = 4 \times 10^{15}$ cm^{-2}) exhibiting characteristic distortion and reduction in channel charge injection. This suggests that the argon-induced donor mechanism is reaching a saturation level, above which the added defects degrade device behavior. Note that the 2ϕ dose characteristics were observed on SA-TG devices that exhibited shifted characteristics as previously described (5), and thus were not included to avoid misperception.

Additional insight is provided by the relatively low temperature thermal stability test at 175 °C. The current degradation as shown in Figure 2a suggests "deactivation" due to instability of the created donor states. The Gate Induced Drain Leakage (GIDL) behavior is pronounced suggesting the creation of secondary defects, which may also degrade on-state charge injection. The anneal-last approach shown in Figure 2b indicates a significant drop in current drive while maintaining low leakage, showing little influence of the argon implant and the associated donor mechanism operative in the implant-last approach.

The results of the boron implanted SA-BG TFTs using the implant-last approach supports the existence of an electrically active donor species involving a relatively small fraction of the total boron concentration; estimated to be less than 10 percent. Devices which received the higher boron dose demonstrated a decrease in current drive that was pronounced at the low drain bias condition as shown in Figure 3b, suggesting an electrically active limit and degradation due to additional boron interstitials and/or other associated defects.

The standard non-SA devices fabricated on the same substrate did not exhibit any left shift upon thermal stability testing. This "active" species appears to be stable at 175 °C, however the balance of boron interstitials appears to be exceedingly mobile and diffuses throughout the channel region. As interstitial boron enters the channel region, a fraction becomes electrically active and shifts the transfer characteristics to the left as shown in Figure 4a. Note that the effective channel length is not changed as shown by the overlay plot in Figure 4b, confirming the stability of the active donor species which is at a much higher concentration in the implanted S/D regions in comparison to the amount which eventually becomes active in the channel. This proposed mechanism is diffusion limited, and thus the degree of shift is channel length dependent as shown in Figure 4c. This may also be the origin of the left-shifted characteristics of the L = 6 μm device shown in Figure 3a, which has a much shorter effective channel length and could experience thermal diffusion during the implant due to heat generated.

The results of the boron implanted SA-BG TFTs using anneal-last approach shows an extensive loss of electrical activity, indicating a nearly complete elimination of the donor species as shown in Figure 5. The passivation anneal is performed for 3 hours at 400 °C, which may allow boron to join the amorphous oxide matrix with a different bonding arrangement which does not support donor-like behavior.

Summary

The interpretation of donor activation in SA-BG IGZO TFTs with ion-implantation of boron ($^{11}B^+$) and argon ($^{40}Ar^+$) species as the source/drain treatment has been presented. The analysis focused on experimental observations of electrical characteristics, with a detailed discussion that developed supporting arguments. The claim of a defect-induced mechanism associated with argon implantation is not surprising, as it is a chemically inert species. However, the interpretation on the behavior of boron in IGZO is unique and certainly not obvious. Results suggest there are at least three distinct states in which boron can exist; two which involve chemical bonding and one as an electrically inactive interstitial atom. Of the two states that involve chemical bonding, only one is proposed to behave as an electrically active donor species. The interstitial diffusivity of boron in IGZO is proposed to be very high, where it appears to move several microns within an hour at 175 °C. In contrast, the electrically active donor state is thermally stable and does not appear to diffuse at such a low temperature. There also appears to be a condition of equilibrium in the ratio of electrically active to inactive species, estimated to be less than 1:9. Materials analysis is needed to provide additional evidence in support of the arguments presented, however the need to identify bonding and ionization states of species at less than one percent presents a tremendous challenge. Comparisons of XPS spectra may be useful in this regard; initial attempts have proven to be problematic. Additional XPS analysis is in progress. ToF-SIMS analysis will also be done for accurate vertical and lateral species profiling.

Acknowledgments

The authors would like to acknowledge the support of the technical staff at the Semiconductor & Microsystems Fabrication Laboratory at RIT, the technical staff at Corning Incorporated, and valuable contributions from supporting research group members. Financial support has been provided by Corning Incorporated and NYSTAR, through the New York State Center for Advanced Technology.

References

1. C. H. Wu, H. H. Hsieh, C. W. Chien, and C. C. Wu, *J. Disp. Technol.*, **5** (12), 515 (2009).
2. D. H. Kang, I. Kang, S. H. Ryu, and J. Jang, *IEEE Electron Device Lett.*, **32** (10), 1385 (2011).
3. J. Park, I. Song, S. Kim, S. Kim, C. Kim, J. Lee, H. Lee, E. Lee, H. Yin, K. Kim, K. Kwon, and Y. Park, *Appl. Phys. Lett.*, **93** (5), 053501 (2008).
4. R. Chen, W. Zhou, M. Zhang, M. Wong, and H. S. Kwok, *IEEE Electron Device Lett.*, **34** (1), 60 (2013).
5. Chowdhury, R. R., Kabir, M. S., Manley, R. G., & Hirschman, K. D., *ECS Transactions*, **92** (4), 135 (2019).
6. K. Terada and H. Muta, *Jpn. J. Appl. Phys.*, **18**, 953 (1979).

Low-Temperature Processed Metal-Semiconductor Field-Effect Transistor with In–Ga–Zn–O Channel Deposited By Ar+O₂+H₂ Sputtering

Y. Magari[a,b] and M. Furuta[a,b]

[a] School of Environmental Science and Engineering, Kochi University of Technology, Kami, Kochi 782-8502, Japan
[b] Center for Nanotechnology, Research Institute, Kochi University of Technology, Kami, Kochi 782-8502, Japan

We investigated the electrical properties of metal-semiconductor field-effect transistors (MES-FETs) with a stacked channel consisting of hydrogenated In–Ga–Zn–O (IGZO:H) on conventional IGZO (IGZO). The IGZO and IGZO:H films were deposited by Ar+O₂ and Ar+O₂+H₂ sputtering, respectively. The maximum process temperature is 150 °C. By applying the IGZO:H/IGZO stacked channel, the on-current of the MES-FET drastically improved compared to the IGZO single-channel, and it exhibited the on-off current ratio of 4.2×10^8 and subthreshold swing of 155 mV/decade. It was found that diffused hydrogen from IGZO:H to IGZO layer increased electrons up to 1.5×10^{19} cm^{-3} near the IGZO:H/IGZO interface. Moreover, the increased electrons confined at the IGZO:H/IGZO interface to form a pseud two-dimensional electron gas because the conduction band offset of 0.18 eV is formed at the IGZO:H/IGZO interface. Thus, the presence of both of the conduction band offset and hydrogen diffusion plays an important role in the MES-FETs performance.

Introduction

Amorphous oxide semiconductors, particularly In–Ga–Zn–O (IGZO) is a promising material for flexible and transparent devices due to their superior properties such as high electron mobility (> 10 cm^{2}V^{-1}s^{-1}), high optical transparency, and low-processing temperature (1,2). To date, extensive efforts have been made to develop IGZO-based metal-insulator-semiconductor field-effect transistors (MIS-FETs) (3,4). In contrast, very few studies have been reported on IGZO-based metal-semiconductor field-effect transistors (MES-FETs) despite the several advantages, such as a simple fabrication process, a low processing temperature, and a low operating voltage due to the lack of a gate insulator (5–9). In order to control the threshold voltage for MES-FETs, the carrier concentration in the channel should be high (~ 10^{18} cm^{-3}); however, it leads to degrading a Schottky contact between the gate electrode and channel (10).

We previously reported that the heterojunction channels consisting of different compositions of IGZO improve the IGZO MIS-FETs properties (11,12). In addition, post-annealing temperature for defect reduction of IGZO MIS-FETs can be reduced from 300 to 150 °C by adding hydrogen gas during the IGZO sputtering (13,14).

In this study, we investigated the electrical properties of MES-FETs with a stacked channel at a maximum process temperature of 150 °C. Two types of stacked channels were examined; stacked IGZO films with different compositions, and stacked IGZO and hydrogenated IGZO films.

Experiment

A top-gate and coplanar MES-FET was fabricated on a glass substrate as shown in Fig. 1(a, b). First, IGZO, hydrogenated IGZO (IGZO:H), In-rich IGZO (HI-IGZO), IGZO/HI-IGZO, and IGZO:H/IGZO were deposited by magnetron sputtering without substrate heating. The IGZO and IGZO:H films were prepared by Ar+O_2 and Ar+O_2+H_2 sputtering from a ceramic $InGaZnO_4$ (In:Ga:Zn = 1:1:1 atom.%) target, respectively. The O_2 and H_2 gas ratios were defined as R[O_2] = O_2/(Ar+O_2+H_2) and R[H_2] = H_2/(Ar+O_2+H_2), respectively. IGZO and IGZO:H were deposited at R[H_2] = 0% and R[H_2] = 5%, respectively, with R[O_2] fixed at 1%. The HI-IGZO film was prepared by Ar+O_2 sputtering from a ceramic $InGaZnO_4$ (In:Ga:Zn = 6:2:1 atom.%) target with an R[O_2] of 30%. As shown in Fig. 1(c), the total channel thickness was maintained at 50 nm for all samples, while the thickness of the stacked channel was 25 nm/25 nm. After patterning of the channels by photolithography and wet etching, the films were annealed in air at 150 °C for 1 h. Then, a 120-nm-thick silver oxide for the Shcottky gate was deposited by reactive sputtering of the Ag target in Ar and O_2 gases with an R[O_2] of 4%, and Au was deposited by vacuum evaporation. Note that resistivity of the Ag_xO film is 3.8×10^{-5} $\Omega \cdot$cm, and thus it can be regarded as a metal (15). Finally, the source and drain electrodes were formed by Mo. The Schottky gate, source, and drain electrodes were patterned by photolithography and lift-off procedures. The channel length (L) and width (W) were 10 and 100 μm, respectively. The device characteristics were measured by a semiconductor parameter analyzer at room temperature (RT) in the dark. Carrier concentration (N_e) of the IGZO films was determined by Hall effect measurements using Van der Pauw geometry at RT. Capacitance–Voltage (C–V) measurements were conducted using an impedance analyzer. The optical properties of the IGZO films were measured by a spectrophotometer and photoelectron yield spectroscopy (PYS). The hydrogen profile in the IGZO stacked film was evaluated using secondary ion mass spectrometry (SIMS).

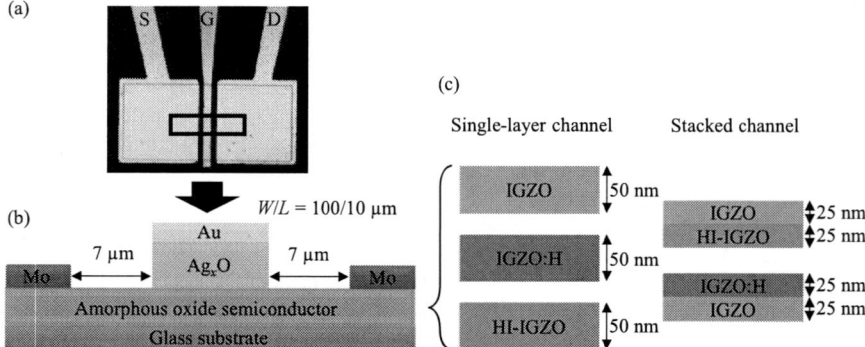

Figure 1. (a) Top-view of fabricated MES-FET (W/L = 100/10 μm), (b) Schematic cross-sectional view of the IGZO MES-FET, and (c) Two types of the IGZO channels used in the experiments.

Results and discussion

Figure 2(a) shows the Tauc plots of the IGZO, IGZO:H, and HI-IGZO films. The optical band gap energy (E_g) of the IGZO, IGZO:H, and HI-IGZO was estimated at 3.04, 3.22, and 2.68 eV, respectively. No noticeable difference was observed in an ionization potential (IP) of each film as shown in Fig. 2(b). From these results, energy band diagrams of the IGZO/HI-IGZO and IGZO:H/IGZO stacked-layer was drawn in Fig. 2(c), suggesting that a conduction band offset (ΔE_c) of 0.36 and 0.18 eV might be formed at the IGZO/HI-IGZO and IGZO:H/IGZO interface, respectively.

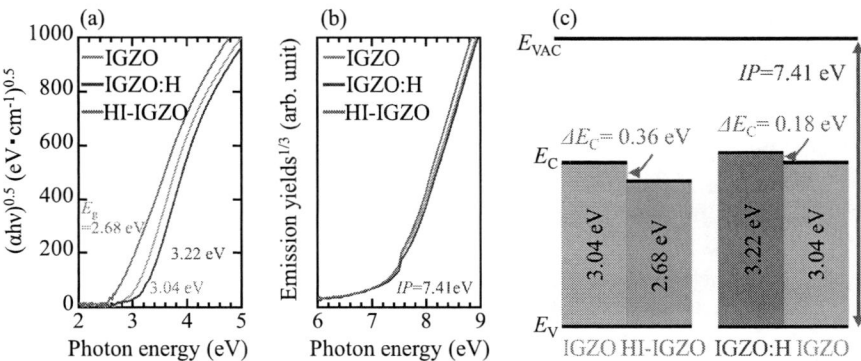

Figure 2. (a) Tauc plots of optical absorption spectra, and (b) PYS spectra of the IGZO, IGZO:H, and HI-IGZO films. (c) Energy band diagrams of the IGZO/HI-IGZO and IGZO:H/IGZO.

Figure 3(a) shows the transfer characteristics of the MES-FET with IGZO and IGZO:H single-channel, and IGZO/HI-IGZO stacked channel. The MES-FETs with IGZO showed switching properties with drain conductance (g_D) of 1.3×10^{-5} S, turn-on voltage (V_{on}) of -0.4 V, subthreshold swing ($S.S.$) of 80 mVdec.$^{-1}$, and on-off current ratio ($I_{on/off}$) of 3.3×10^7 as summarized in TABLE I. In contrast, the MES-FET with HI-IGZO did not show switching properties (conductive behavior), which is possibly due to gate leakage current at the Ag$_x$O/HI-IGZO Schottky interface resulting from the large electron affinity of HI-IGZO film. MES-FET with IGZO/HI-IGZO stacked channel showed similar characteristics to the MES-FET with the IGZO single-channel. Although the ΔE_c of 0.36 eV was expected at the IGZO/HI-IGZO interface, an improvement of the device performance was not observed for the MES-FET with the IGZO/HI-IGZO channel.

Figure 3(b) shows the transfer characteristics of the MES-FET with IGZO and IGZO:H single-channel, and IGZO:H/IGZO stacked channel. The MES-FET with IGZO:H showed positive V_{on} and low on-current (I_{on}) compared to the IGZO channel. This is due to a low carrier concentration (N_e) of the IGZO:H film ($N_e = 4.4 \times 10^{17}$ cm^{-3}) than that of the IGZO film ($N_e = 7.6 \times 10^{17}$ cm^{-3}) as shown in TABLE I. By stacking the IGZO:H on the IGZO, N_e of 1.4×10^{19} cm^{-3} was observed for the IGZO:H/IGZO stacked film which is over an order of magnitude higher than those of IGZO and IGZO:H single layer. Moreover, the MES-FET with IGZO:H/IGZO stacked channel exhibited the superior switching properties with a g_D of 1.1×10^{-4} S, $I_{on/off}$ of 4.2×10^8, V_{on} of -5.9 V, $S.S.$ of 155 mVdec.$^{-1}$, and low-off current of 7.4×10^{-13} A. In particular, obtained $I_{on/off}$ was higher than those of reported amorphous oxide semiconductor MES-FETs, as summarized in Table II. Unfortunately, field-effect mobility (μ_{FE}) could not be compared because the μ_{FE} of the MES-FETs is calculated assuming a constant N_e in the whole channel (16). We successfully demonstrated the high-performance MES-FET characteristics by employing a stacked channel consisting of IGZO:H/IGZO.

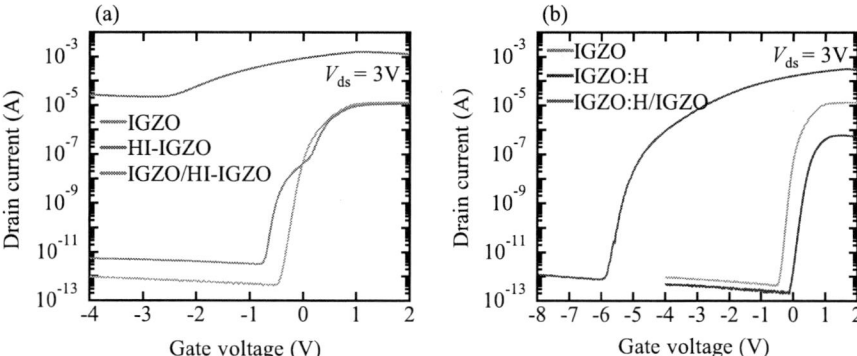

Figure 3. Transfer characteristics of the MES-FETs with (a) IGZO, HI-IGZO, and IGZO/HI-IGZO channels, and (b) with IGZO, IGZO:H, and IGZO:H/IGZO channels.

TABLE I. Summary of electrical properties of the MES-FETs with various channel materials.

Channel material	IGZO	HI-IGZO	IGZO/ HI-IGZO	IGZO:H	IGZO:H/ IGZO
g_D (S)	1.3×10^{-5}	-	9.1×10^{-6}	2.0×10^{-8}	1.1×10^{-4}
V_{on} (V)	-0.4	-	-0.6	-0.1	-5.9
S.S. (mVdec.$^{-1}$)	80	-	100	107	155
$I_{on/off}$	3.3×10^{7}	-	3.7×10^{6}	3.2×10^{6}	4.2×10^{8}
N_e (cm^{-3})	7.6×10^{17}	2.2×10^{16}	7.0×10^{16}	4.4×10^{17}	1.4×10^{19}

TABLE II. Survey of characteristics of amorphous oxide semiconductor MES-FETs.

Refs.	Gate/channel	W/L (μm)	μ_{FE} (cm^{2}V^{-1}s^{-1})	S.S. (mVdec.$^{-1}$)	$I_{on/off}$	T_{max} (°C)
This work	Ag$_x$O/IGZO:H/IGZO	100/10	-	155	4.2×10^{8}	150
(17)	Ag$_x$O/IGZO	430/10	14.1	112	2.5×10^{7}	150
(5)	Ag$_x$O/IGZO	785/5	3.2	356	3.8×10^{7}	350
(8)	RuSiO$_x$/IGZO	250/10	9	250	2×10^{5}	RT
(10)	Pt/IGZO	300/100	0.5	129	$\sim10^{7}$	200
(18)	PtO$_x$/ZTO	200/3	5	242	$\sim10^{8}$	110
(19)	PtO$_x$/ZTO	430/10	0.9	124	1.8×10^{6}	RT
(20)	Ag$_x$O/ZTO	260/10	12	180	8×10^{6}	525
(21)	Ag$_x$O/SnO$_2$	262/10	0.33	207	7.3×10^{6}	350
(22)	Ag$_x$O/Ga$_2$O$_3$	524/5	1.3	-	2×10^{7}	400
(23)	Ag$_x$O/ZnO	100/5	8	140	$\sim10^{6}$	800
(24)	PtO$_x$/ZnO	10.75	9.5	130	1×10^{5}	700
(25)	Ag$_x$O/ZnMgO	10.75	11.9	-	1.4×10^{6}	670
(26)	Ag$_x$O/ZnO	50/10	15	-	$\sim10^{4}$	800
(27)	Ag$_x$O/ZnO	10.75	1.3	125	4.7×10^{5}	675
(28)	Ag$_x$O/ZnO	400/60	11.3	-	2×10^{8}	630

To investigate the enhancement mechanisms of the IGZO:H/IGZO MES-FET, the depth profiles of hydrogen and background charge density (N_{bg}) in the IGZO:H/IGZO stacked film were extracted by SIMS and C–V measurements, respectively. It is well known that the hydrogen acts as a shallow donor in the oxide semiconductor (2). Figure 4(a) shows the depth dependence of hydrogen ion intensity and N_{bg} of the IGZO:H/IGZO stacked film after annealing at 150 °C. It was confirmed that the hydrogen in the IGZO:H layer was over an order of magnitude higher than that in the IGZO layer. Moreover, it was found that such hydrogen atoms diffused from the IGZO:H layer to the IGZO layer up to about 40 nm.

The depth profile of N_{bg} in the IGZO:H/IGZO stacked film is also shown in Fig. 4(a). A vertical structure Schottky diode was fabricated and used for C–V analysis. The N_{bg} in the IGZO:H layer was ~4×10^{17} cm^{-3}, which is consistent with the N_e of IGZO:H single layer obtained by Hall effect measurement as shown in TABLE I. For the IGZO layer, an extremely large N_{bg} of 1.5×10^{19} cm^{-3} was observed at a depth of 26 nm, and the N_{bg} gradually decreased with increasing depth. The obtained N_{bg} (1.5×10^{19} cm^{-3}) at the IGZO:H/IGZO interface was an order of magnitude higher than N_e of IGZO single layer (7.6×10^{17} cm^{-3}), whereas the N_{bg} is in good agreement with the N_e of IGZO:H/IGZO stacked film (1.4×10^{19} cm^{-3}) as shown in TABLE I. It was thus found that the high-I_{on} in IGZO:H/IGZO MES-FET is due to the increased free electrons at the IGZO:H/IGZO interface.

Here, we discuss the origin of high-N_{bg} at IGZO:H/IGZO interface. We previously reported that hydrogen in the IGZO:H acts as a carrier suppressor (13,14), whereas diffused hydrogen in the IGZO acts as a shallow donor (29). The decay curve of H ion intensity and N_{bg} as a function of depth showed a similar tendency (Fig. 4(a)). Thus, it is reasonable to consider that diffused hydrogen in the IGZO layer increased N_{bg} near the IGZO:H/IGZO interface. As shown in Fig. 4(b), the electrons in the IGZO layer generated by the hydrogen diffusion would be effectively confined at the IGZO:H/IGZO interface to form a pseud two-dimensional electron gas (2DEG) because ΔE_c of 0.18 eV is formed at the IGZO:H/IGZO interface (Fig. 2(c)). As a result, high-I_{on} and low-I_{off} could be achieved simultaneously in the IGZO:H/IGZO MES-FET.

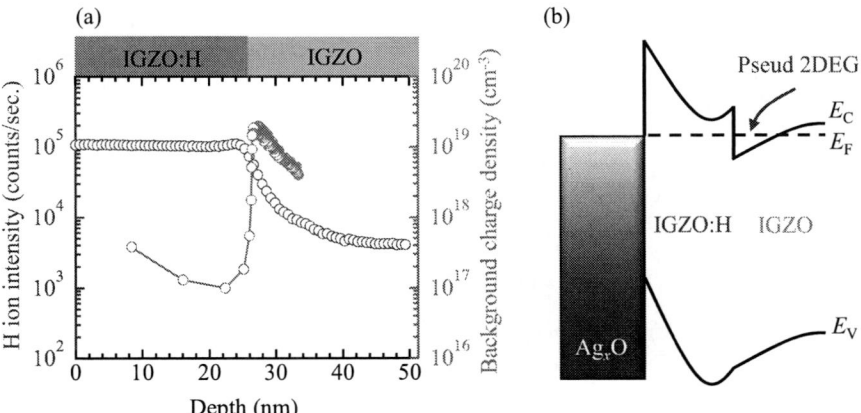

Figure 4. (a) Depth profiles of hydrogen and N_{bg} in the IGZO:H/IGZO stacked film after annealing at 150 °C. (b) Schematic energy band diagram just below the gate electrode of MES-FET with IGZO:H/IGZO stacked channel.

Conclusions

In this study, stacked channel MES-FETs consisting of IGZO/HI-IGZO and IGZO:H/IGZO were fabricated at a low temperature of 150 °C. In the case of MES-FET with an IGZO/HI-IGZO stacked channel, although the ΔE_c of 0.36 eV was expected at the IGZO/HI-IGZO interface, an improvement of the device performance was not observed. In contrast, by applying the IGZO:H/IGZO stacked channel, the I_{on} of the MES-FET drastically improved compared to the IGZO single-channel while maintaining a low-I_{off}. The $I_{on/off}$ of 4.2×10^8, V_{on} of -5.9 V, and $S.S.$ of 155 mVdec.$^{-1}$ were achieved. To the best of our knowledge, obtained $I_{on/off}$ is the highest value among the amorphous oxide semiconductor MES-FETs produced at various process temperatures (RT–800 °C) reported to date. SIMS and C–V analysis revealed that diffused hydrogen from IGZO:H to IGZO layer increased electrons up to 1.5×10^{19} cm^{-3} near the IGZO:H/IGZO interface. Moreover, increased electrons would be effectively confined at the IGZO:H/IGZO interface to form a pseud 2DEG because ΔE_c of 0.18 eV is formed at the IGZO:H/IGZO interface. Thus, the presence of both of the ΔE_c and hydrogen diffusion plays an important role in the MES-FETs performance. The proposed method successfully demonstrated great potential for future flexible electronics applications.

Acknowledgments

The SIMS measurement was supported by "Nanotechnology Platform" (project No.S-19-NR-0015) of the Ministry of Education, Culture, Sports, Science and Technology (MEXT), Japan.

References

1. K. Nomura, H. Ohta, A. Takagi, T. Kamiya, M. Hirano, and H. Hosono, Nature **432**, 488 (2004).
2. T. Kamiya, K. Nomura, and H. Hosono, J. Disp. Technol. **5**, 273 (2009).
3. T. Toda, G. Tatsuoka, Y. Magari, and M. Furuta, IEEE Electron Device Lett. **37**, 1006 (2016).
4. S.G.M. Aman, D. Koretomo, Y. Magari, and M. Furuta, IEEE Trans. Electron Devices **65**, 3257 (2018).
5. G.T. Dang, T. Kawaharamura, M. Furuta, and M.W. Allen, IEEE Electron Device Lett. **36**, 463 (2015).
6. Y. Magari, S. Hashimoto, K. Hamada, and M. Furuta, ECS. Trans. **75**(10), 139 (2016).
7. M. Furuta, Y. Magari, S. Hashimoto, and K. Hamada, ECS. Trans. **79**(1), 43 (2017).
8. J. Kaczmarski, S. Member, A. Taube, M.A. Borysiewicz, M. Mysliwiec, K. Piskorski, K. Stiller, and E. Kaminska, IEEE Trans. Electron Devices **65**, 129 (2018).
9. Y. Magari, S.G.M. Aman, D. Koretomo, K. Masuda, K. Shimpo, and M. Furuta, Jpn. J. Appl. Phys. **59**, SGGJ04 (2019).
10. D.H. Lee, K. Nomura, T. Kamiya, and H. Hosono, ECS Solid State Lett. **1**, Q8 (2012).

11. M. Furuta, D. Koretomo, Y. Magari, S.G.M. Aman, R. Higashi, and S. Hamada, Jpn. J. Appl. Phys. **58**, 090604 (2019).
12. D. Koretomo, S. Hamada, Y. Magari, and M. Furuta, Materials **13**, 1935 (2020).
13. S.G.M. Aman, Y. Magari, K. Shimpo, Y. Hirota, H. Makino, D. Koretomo, and M. Furuta, Appl. Phys. Express **11**, 081101 (2018).
14. D. Koretomo, S. Hamada, M. Mori, Y. Magari, and M. Furuta, Appl. Phys. Express **13**, 076501 (2020).
15. Y. Magari, H. Makino, S. Hashimoto, and M. Furuta, Appl. Surf. Sci. **512**, 144519 (2019).
16. S. M. Sze, Physics of Semiconductor Devices 2nd ed. Chap, Wiley Interscience, New York (1981).
17. M. Lorenz, A. Lajn, H. Frenzel, H. V. Wenckstern, M. Grundmann, P. Barquinha, R. Martins, and E. Fortunato, Appl. Phys. Lett. **97**, 243506 (2010).
18. O. Lahr, Z. Zhang, F. Grotjahn, P. Schlupp, S. Vogt, H. Von Wenckstern, A. Thiede, and M. Grundmann, IEEE Trans. Electron Devices **66**, 3376 (2019).
19. S. Vogt, H. Von Wenckstern, and M. Grundmann, Appl. Phys. Lett. **113**, 133501 (2018).
20. G.T. Dang, T. Kawaharamura, M. Furuta, and M.W. Allen, Appl. Phys. Lett. **110**, 073502 (2017).
21. G.T. Dang, T. Uchida, T. Kawaharamura, M. Furuta, A.R. Hyndman, R. Martinez, S. Fujita, R.J. Reeves, and M.W. Allen, Appl. Phys. Express **9**, 041101 (2016).
22. G.T. Dang, T. Kawaharamura, M. Furuta, and M.W. Allen, IEEE Trans. Electron Devices **62**, 3640 (2015).
23. S. Elzwawi, A. Hyland, M. Lynam, J.G. Partridge, D.G. McCulloch, and M.W. Allen, Semicond. Sci. Technol. **30**, 24008 (2015).
24. F.J. Klüpfel, F. Schein, M. Lorenz, H. Frenzel, H. Von Wenckstern, and M. Grundmann, IEEE Trans. Electron Devices **60**, 1828 (2013).
25. H. Frenzel, A. Lajn, H. Von Wenckstern, and M. Grundmann, J. Appl. Phys. **107**, 114515 (2010).
26. S. Elzwawi, H.S. Kim, M. Lynam, E.L.H. Mayes, D.G. McCulloch, M.W. Allen, and J.G. Partridge, Appl. Phys. Lett. **101**, 243508 (2012).
27. H. Frenzel, M. Lorenz, A. Lajn, H. Von Wenckstern, G. Biehne, H. Hochmuth, and M. Grundmann, Appl. Phys. Lett. **95**, 153503 (2009).
28. H. Frenzel, A. Lajn, M. Brandt, H. Von Wenckstern, G. Biehne, H. Hochmuth, M. Lorenz, and M. Grundmann, Appl. Phys. Lett. **92**, 192108 (2008).
29. T. Toda, Deapeng Wang, Jingxin Jiang, Mai Phi Hung, and M. Furuta, IEEE Trans. Electron Devices **61**, 3762 (2014).

Improved Copper Electrode Integration for Thin Film Electronics on Glass

Hoon Kim, Bin Zhu, Rajesh Vaddi, Ming-Huang Huang, Robert G. Manley

Corning Research and Development Corporation, 1 Riverfront Plaza, Corning, NY, USA

> We demonstrate single layer Cu interconnects using thin CuMn alloy as a temporary adhesion layer. After short annealing at 300 °C for 60 s of a 10 nm CuMn alloy with 500 nm of Cu film, CuMn alloy layer is converted to pure Cu and the Mn is reacted with glass to form MnO_x serving as an adhesion layer. After optimization of Mn concentration, alloy thickness, annealing environment, time and temperature, a resistivity lower than 1.8 $\mu\Omega$·cm of the electrode was achieved. We also evaluated the effect of glass surface treatment to confirm the sensitivity of this process depending on substrate preparation conditions. This process has been shown to be very stable for different acid treatments on display glasses. It is also compared to a Ti/Cu stack as a reference.

Introduction

As display sizes become larger, copper is the leading electrode conductor material due to its low resistivity and high reliability (1). Currently, on oxide glass substrates, an adhesion layer, such as Ti, Ta, Mn or Mo, is required for Cu to prevent delamination and increase reliability. However, this adhesion layer impacts attributes such as an increase in the resistivity of Cu and more complicated two step etching process for patterning electrodes (2). To address this issue, alloying of Cu with Mg, Mg/Al, Ti, Cr, Zr, Mn, Hf, or Ru are used to form a diffusion and adhesion layer (2~9). However, these alloys require long annealing times for the out diffusion of the alloy elements, and lead to higher resistivity than that of pure Cu due to some residual elements in the grain boundary or bulk of the film. CuMg/Cu bilayer approach was evaluated to address resistivity issue (10). However, the CuMg/Cu layer stack has a high resistivity, 2.63 $\mu\Omega$·cm due to the thick 50 nm of CuMg adhesion layer and high concentration of Mg at 4.5 at% in Cu. CuMn alloy demonstrates a potential of low resistivity which has a self-forming barrier by reaction with SiO2. The resistivity of the alloy can be reduced to <2.0 $\mu\Omega$·cm in diluted oxygen environment (11). However, such an annealing environment for Mn out-diffusion needs an optimize time to minimize the oxidation of the Cu film but be long enough time for Mn out diffusion of a thick film (11). In this paper, we take advantage of Mn as an alloy element which enables low resistivity Cu alloy and improve the Cu adhesion on glass substrates. We proposed thin CuMn alloy, <10 nm, as an adhesion layer on a glass instead of a thick full alloy stack to reduce the annealing time for Mn out diffusion and for low resistivity to minimize Mn incorporation in Cu film on different treated display glass. This thin alloy layer converts to a pure Cu layer after annealing by reaction of Mn and reaction with glass.

The stack had a resistivity of <1.8 $\mu\Omega\cdot$cm with good adhesion of >3 N/cm by forming an intermixing layer of Cu and glass with Mn. There is no detectable difference between the surface treatments with various acid cleaning used in display integration processes. Thus, it is a promising thin-film Cu electrode integration methodology adhesion on glass and low resistivity with a simplified single etch process.

Experimental

Coring® Eagle XG® glass is used as a substrate which is introduced into an e-beam evaporation chamber for CuMn alloy deposition. Mn concentration is varied from 0.5 at% to 2 at%, and the alloy thickness is from 10 nm to 50 nm depending on condition. The alloy pallets were prepared by Kurt J. Lesker. Pure Cu was deposited by the e-beam evaporation on the CuMn at 50 °C with and without vacuum break between the CuMn and pure Cu deposition to determine the effect oxidation of the CuMn layer. The film thickness was monitored in situ with quartz crystal microbalance (QCM). Figure 1 shows the sample structure. Layer 1 is CuMn alloy on glass acting as the adhesion layer and layer 2 is a pure Cu as making up the majority of the electrode conductor. We evaluated the post annealing effects under N_2 and forming gas (FGA: 4% H_2 in N_2) environments at temperature from 300 °C to 350 °C at times between 60 s to 2000 s to compare the annealing condition effect in a rapid thermal annealing chamber at atmospheric pressure. Cu/CuMn stack was compared with Cu/Ti stack in terms of resistance change depending on annealing and Ti or Mn diffusion to thick Cu layer. Ti film was deposited using a sputtering process, and sputtering Cu was deposited on Ti without vacuum break using cluster PVD tool to avoid the oxidation of Ti surface. CuMn is deposited at the evaporator and move to the sputtering tool for Cu deposition, to use same Cu deposition method for fair comparison. Sheet resistance is measured using the four-point probe method for with resistivity being extracted using the measure film thicknesses for each annealing condition. The resistivity of electrode is calculated based on the sum of thickness of the CuMn and Cu films. Adhesion is evaluated with the 3 N tape test with cross-hatches based on the ASTM D3359-09e2 standard before and after annealing. TEM cross-section with energy dispersive spectrometry (EDS) mapping confirmed the formation of interfaces layers and the thickness of each film. To evaluate the surface treatment effect on the adhesion of CuMn alloy on glass, it is treated with HCl, nitric acid, sulfuric acid, and compared with no acid treated glass. Adhesion and resistivity of the stack were compared.

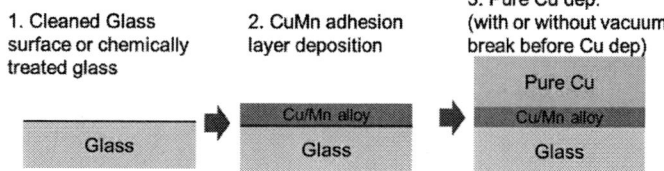

Figure 1. Process flow and CuMn/Cu structure. A thin layer CuMn is deposited on surface treated or cleaned glass substrate followed by the deposition of a thicker pure Cu layer.

Results and Discussion

The effect of Mn concentration in the CuMn alloy on resistivity of the stack was evaluated depending on annealing time shown in Figure 2. With decreasing Mn concentration, the resistivity of Cu film stack also decreases. The resistivity appears to reach a lower limit with longer annealing times with the range considered. The concentration of Mn in the thin 10 nm alloy layer impacts the total resistivity of the film stack. More detail of the origin of this result is discussed further in the TEM analysis.

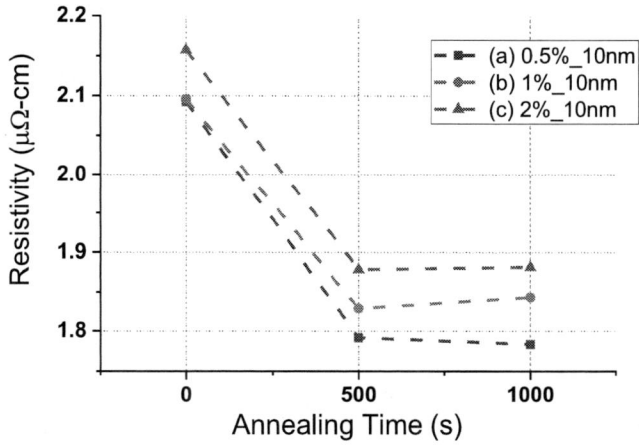

Figure 2. Resistivity of the film stack as a function of annealing time in FGA at 300 °C. Mn concentrations include (a) 0.5, (b) 1, (c) 2 at% for 10 nm CuMn adhesion layer and 500 nm Cu film.

To evaluate the thickness effect of the alloy layer, films of 10, 20, and 50 nm are used with Mn concentration 0.5 at% and 1 at%. Thinner alloy layers with lower Mn concentration had lower resistivity, shown in Figure 3. A film with 0.5 at% of Mn in 10 nm thick alloy layer stack showed the resistivity <1.8 μΩ·cm which is almost same as a pure Cu prepared in house sample. Thicker alloy layers with high Mn concentration required longer annealing times to obtain the lower resistivity levels. This is because the out diffusion of Mn atoms takes longer time for thicker films and higher concentrations. Alloy films of 20 nm with 0.5 at% Mn and 10 nm films with 1 at% Mn result in CuMn alloy films having the same total amount of manganese in the layer which have the same resistivity. Thus, the total amount of Mn in the stack determines the resistivity of the stack.

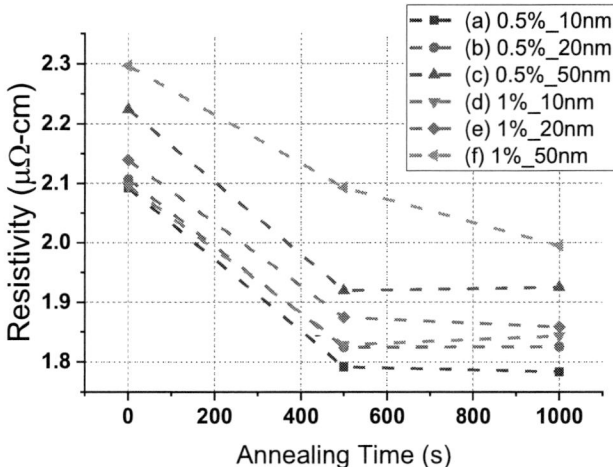

Figure 3. Resistivity of the film stacks (500 nm Cu/ CuMn alloy) as a function of annealing time (FGA at 300 °C) for different Mn concentrations of (a)~(c) 0.5 at% and (d)~(f) 1 at% Mn. Thickness is varied between 10~50 nm for the CuMn adhesion layer.

Figure 4. The adhesion test result using 3 N tape with cross hatch for 10 nm 0.5 at% Mn in Cu with 500 nm Cu stack for (a) as-deposited and (b) after 300 °C, 500 s anneal.

Although the as-deposited alloy samples failed the adhesion test, shown in Figure 4 (a), all the alloy samples, even with 10 nm of 0.5 at% Mn alloy layer passed the 3 N tape test after 300 °C annealing (Figure 4. (b)). Thus, such small amount of Mn part of the interface between Cu and glass improved the adhesion of Cu film by intermixing of Mn and glass. Mn diffusion to the glass will be discussed with the TEM results.

(a)

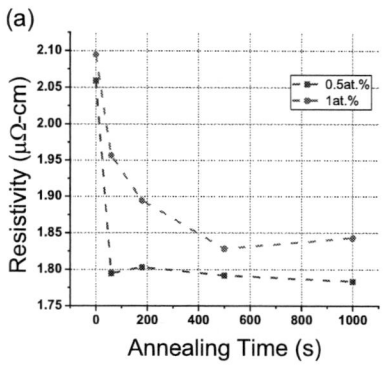

(b)

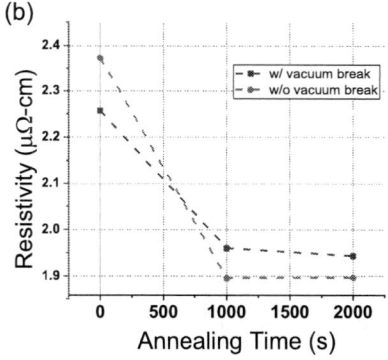

Figure 5. Resistivity of Cu stacks for (a) 10 nm of 0.5 or 1 at% Mn and 500nm Cu, (b) 10 nm 2 at% CuMn with and without vacuum break between alloy and Cu deposition as a function of annealing time at 300 °C under FGA environment.

The resistivity change as a function of annealing time is evaluated using 10 nm of 0.5 and 1 at% Mn alloy as adhesion layer shown in Figure 5. (a). After 60 s annealing at 300 °C in FGA exhibits a saturation of the resistivity minimum which implies Mn takes time to diffuse to the glass interface from the 10 nm alloy and leave behind pure Cu. Since Mn diffusion in Cu is through grain boundary and is about 30 times higher than that of bulk, (12) such fast Mn diffusion to glass is possible through the grain boundary of 10 nm alloy layer. Considering the process of metallization, a degas chamber can be used as an annealing chamber for Mn out-diffusion (13). The impact of a vacuum break between CuMn alloy and pure Cu deposition is considered. It was believed that the oxidation of CuMn alloy surface, by breaking vacuum between CuMn and Cu deposition, an oxide layer would form on the surface. It could act as a barrier layer for Mn diffusion into thick pure Cu and help prevent the increase of resistivity. As shown in Fig. 5 (b), however, no air-break between CuMn and Cu had a lower resistivity compare to that having a vacuum break. Although the Mn diffusion into pure Cu may increase the resistivity, the oxidation of the 10 nm alloy layer also increases the resistivity. Thus, oxidation of the alloy layer appears to lead to a larger increase in resistivity compare to that of Mn diffusion into the pure Cu film. TEM EDS mapping results help to explain the trend of resistivity depending on processing as shown in Figure 6. The Mn signal with vacuum break (Figure 6 (a)) has a narrow distribution compare to that of continuous deposition in vacuum as expected due to stabilization of Mn in the oxidized CuMn alloy. Although the no vacuum break sample has a wide spread of Mn at the interface, most of the Mn is located at the glass interface shown by TEM (Figure 6. (b)). Since glass is the only source of oxygen for the Mn sample in no-vacuum break condition, more intermixing of Mn and glass is observed. Thus, the no-vacuum break sample has less amount of Mn in the pure Cu layer, indicated above of yellow dot line in the in Figure 6(a) and (b), and less electron scattering near the interface. However, the vacuum-break sample has less Mn diffusion to glass interface, and most of Mn located in the Cu due to oxidation of CuMn alloy. Larger amounts of Mn near interface of the alloy and Cu is a scattering center leading to an increase the overall resistivity. This result may explain the previous observations of the high resistivity of Cu film with high concentration of Mn at the adhesion layer. More Mn at the adhesion layer will remain near Cu and glass interface which can be an origin of a resistivity increase becoming a scattering center. Thus, right amount of Mn is critical for low resistivity with good adhesion.

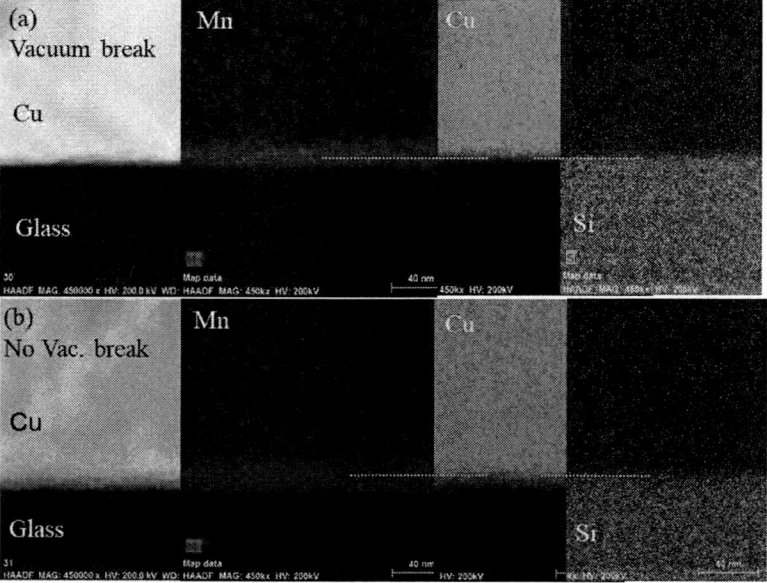

Figure 6. TEM EDS elemental maps for (a) vacuum break between CuMn (2 at%) and 500 nm thick Cu (b) No vacuum break (in-situ deposition) after 300 °C, 2000 s annealing.

Iijima *et. al.* reported the optimized oxygen concentration during annealing to enhance the Mn out-diffusion to achieve lower resistivity (11). The effect of annealing environment on Cu resistivity was evaluated by comparing Ar condition which has <100 ppm O_2 as impurity in an FGA, which is a reducing environment. Both Ar and FGA conditions resulted in the same low resistivity level 1.89 $\mu\Omega\cdot$cm. However, the resistivity of the Ar condition increases with longer annealing time. It is suspected that this is due to the oxidation of the Cu film. FGA at 350 °C also exhibited a resistivity increase at longer annealing times as well. This may be due to the roughness increase by agglomeration of the Cu film annealed at higher temperatures. Thus, forming gas annealing at 300 °C is the best condition for low resistivity and smooth morphology of the 500 nm Cu film.

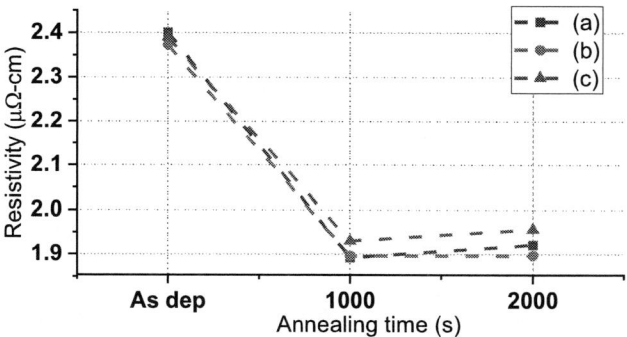

Figure 7. Resistivity of the Cu stacks of 10 nm CuMn alloy with 2 at% Mn / 500 nm Cu: no-vacuum break as a function of time for different annealing conditions of (a) Ar at 300 °C, (b) FGA at 300 °C, and (c) FGA 350 °C.

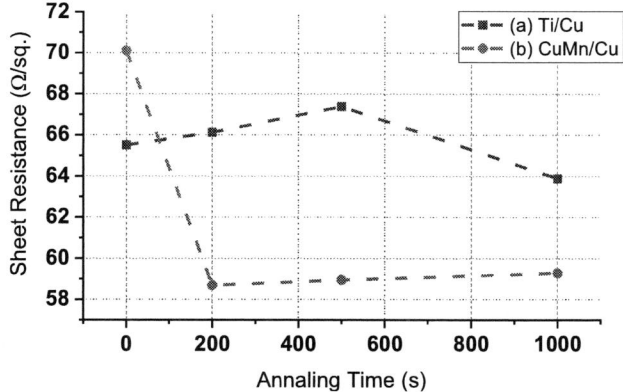

Figure 8. The sheet resistance of (a) Ti/Cu and (b) CuMn/Cu stack on the glass depending on annealing time under FGA environment at 300 °C.

Figure 8 shows the sheet resistance change of (a) Ti (60 nm) /Cu (330 nm) and (b) CuMn 2 at% (10 nm)/ Cu (320 nm) stack. Although the CuMn/Cu stack has higher sheet resistance for the as-deposited condition, it decreases after annealing. However, the resistance of Cu/Ti stack does not decrease after annealing and even the grain size becomes larger comparing Figure 9(a) as-deposited and (c) post annealed cases. The resistance of Cu stack with pure metal adhesion layer is a higher value due to diffusion of metal element in to the pure Cu. Ti has solubility in about 0.5 at% at 400 °C, and the resistivity of Cu (Ti 0.5 at.%) alloy is 3.1 $\mu\Omega \cdot cm$ (14). Thus, if Ti diffuses into Cu during annealing, the resistance of the Cu stack may not decrease even with larger grain size. To confirm this concept, TEM EELS is used to track the Ti diffusion within the Cu conductor.

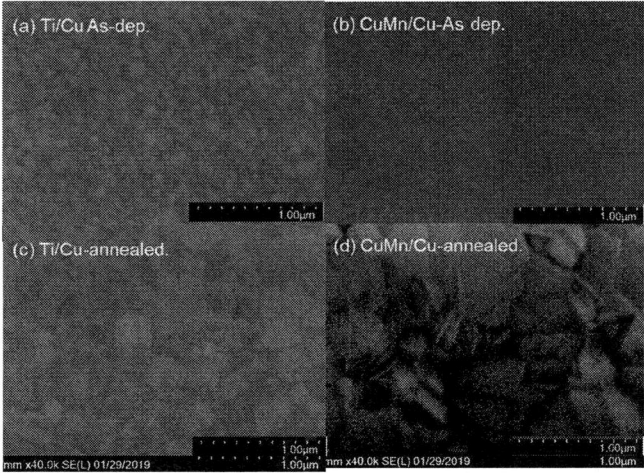

Figure 9. The SEM images of as-deposited sputtering Cu film on (a) Ti, (b) CuMn 2 at%, and post annealing at 300 °C for 1000 s on (c) Ti 20 nm, and (d) CuMn 2 at% 10 nm.

Figure 10 is the TEM EELS mapping Cu Cu/Ti stack after FGA annealing at 300 °C for 500 s. Figure 10 (a) is a TEM image which indicates the grain boundary location of Cu layer. No Ti signal is detected in the grain boundary, but a Ti peak is detected on top of the Cu layer where the grain boundary is in figure 10 (c). This result is a clear evidence that Ti out-diffused through grain boundary of Cu layer. Although Ti peak is not observed in the grain boundary of Cu due to detection limit of the EELS, Ti atoms are supposed to be located throughout grain boundaries. Ti atoms in the grain boundary of Cu act as scattering center for electrons. Thus, why the resistance does not decrease for Cu/Ti stack can be explained with this result. However, for Cu/CuMn stack, no Mn out-diffusion is detected in the TEM EELS analysis in figure 11 (b). Mn is located only near the interface, and no trace of Mn is observed on the top of the Cu layer. Thus, the CuMn alloy seed shows lower resistance compare to that of Cu/Ti stack even thinner thickness.

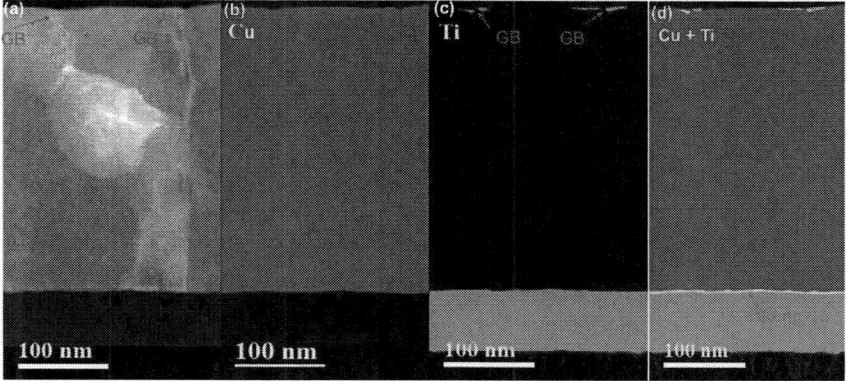

Figure 10. TEM of Cu (332 nm)/ Ti (60 nm) (a) bright field image (b) Cu EELS (c) Ti EELS, and (d) EELS image Cu + Ti mapping after FGA annealing at 300 °C for 500 s.

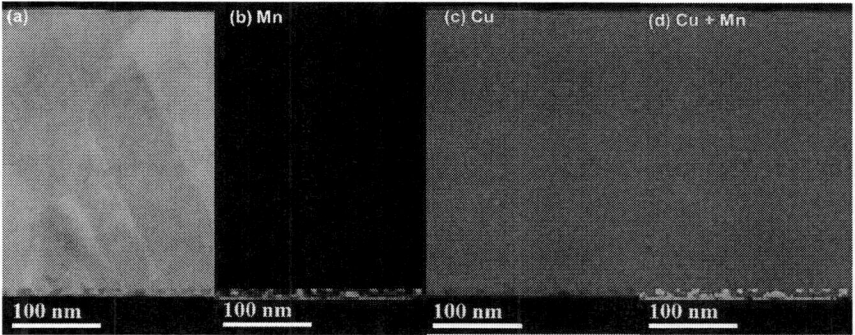

Figure 11. TEM of Cu (322 nm)/ CuMn 2 at% (10 nm) (a) bright field image (b) Mn EELS (b) Cu EELS, and (d) EELS image Cu + Mn mapping after FGA annealing at 300 °C for 500 s.

Considering integration process, the surface cleaning of glass before metallization process may different. Thus, the effect of surface cleaning process on the adhesion of CuMn alloy is evaluated. We compared few the most commonly used acid cleaning method, such as HCl, HNO$_3$, and H$_2$SO$_4$. Normal POR (detergent) cleaning of bare EXG glass is used as a reference. Fig. 12 is the sheet resistance of Cu stack on treated glasses or untreated glass. There are very minor differences between treatments. The adhesion on a treated surface is same as bare EXG glass. The as-deposited film fails the tape test, but it passes the adhesion test after annealing. Thus, this process is applicable for various surface cleaning method depending on manufacturing preference.

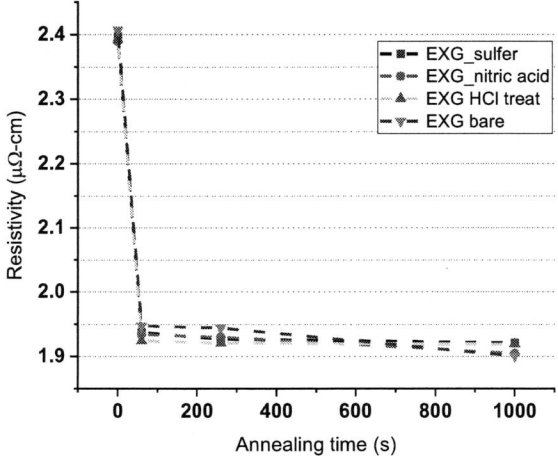

Figure 12. The sheet resistance of Cu stack (350 nm Cu/ 10 nm CuMn 0.5 at%) on treated glasses.

Conclusions

Demonstration of the use of CuMn alloy layer to promote thick Cu conductor film adhesion to Corning® Eagle XG® substrates and result a single layer Cu film is reported. The Mn concentration is evaluated 0.5~2 at% and the thickness of adhesion alloy layer from 10 to 50 nm. Concentrations of 0.5 at% Mn in 10 nm CuMn alloy successfully served as an adhesion layer for 500 nm thick Cu films. No peeling occurred as characterized by the 3 N tape test after 300 °C for 60 s annealing. The resistivity of film stack is almost same as pure Cu at <1.8 µΩ·cm. The forming gas annealing effectively protects the further oxidation of Cu during annealing and no additional oxygen is required to extract the Mn in the Cu film as the glass is a good source of oxygen for the small amount of Mn atoms in the thin alloy layer. TEM EDS confirmed the Mn and glass interdiffusion which improves the adhesion of Cu to the glass. The amount of Mn near the interface affects the resistivity of the film. Thus, the proper amount of Mn in the alloy layer is critical for low resistivity and adhesion. No Mn out-diffusion to thick Cu is detected, but Ti diffuses out to thick Cu through grain-boundary which is confirmed by TEM EELS mapping. It is immune to surface cleaning methods. Thus, it is a promising method for Cu interconnect processes without barrier/adhesion layer for simple single metal integration process and lower line conductance for large display application.

References

1. HIS market technology, "Cu electrode, the key to UHD LCD technology", August 13, (2014)
2. J. Koike, K. H., M. Naito, P. Yun, and Y. Sutou, "P-33: Cu-Mn Electrodes for a-Si TFT and Its Electrical Characteristics" in Proc. SID 2010, p. 1343.(2010)
3. W. H. Lee, H. L. Cho, B. S. Cho, J. Y. Kim, W. J. Nam, Y-S. Kim, W. G. Jung, H. Kwon, J. H. Lee, and J. G. Lee, *Appl. Phys. Lett.* 77, 2192 (2000)
4. W. Lee, H. Cho, B. Cho, J. Kim, Y. Kim, W.Jung, H. Kwon, and J. Lee, *Journal of Vacuum Science & Technology* A 18, 2972 (2000)
5. H. Sirringhaus, S. D. Theiss, A. Kahn, and S. Wagner, *IEEE Electron Device Lett.* 0741-3106, 18, 388 (1997)
6. Z. Yu, J. Xue, Qi Yao, Z. Li, G. Hui, W. Xu, *Microelectronic Engineering* 170, 16-20. (2017)
7. J.S. Fang, Y.T. Chen, *Surface and Coatings Technology* 231, 166-170. (2013)
8. J. H Lee, C. Y. Lee, H. S. Nam, J. G Lee, H. J. Yang, W. J. Ho, J. Y. Jeong, D. H. Koo, *Journal of Electronic Materials* 40:11, 2209-2213. (2011)
9. J.P. Chu, C.H. Lin, and V.S. John, *Appl. Phys. Lett.* 91, 132109 (2007)
10. M. C. Wang, T.-C. Chang, Po-Tsun Liu, Y. Y. Li, R. W. Xiao, L. F. Lin and J. R. Chen, *Electrochem. Solid-State Lett.*, 10, p. J83 (2007)
11. J. Iijima, Y. Fujii, K. Neishi, and J. Koike, *Journal of Vacuum Science & Technology* B: 27, 1963 (2009)
12. Y. Au, Y. Lin, H. Kim, E. Beh, Y. Liu and R. G. Gordon, *J. Electrochem. Soc.* volume 157, issue 6, D341-D345 (2010)
13. K-M. Chang, W-C. Yang, C-P. Tsai, *IEEE Electron Device Letters,* Volume: 24 , Issue: 8 , Aug. (2003)
14. K. Barmak, A. Gungor, C. Cabral Jr., and J. M. E. Harper, *Journal of Applied Physics* 94, 1605 (2003)

Effect of glass substrate on the film properties of poly silicon by excimer laser annealing

Bin Zhu, Rajesh Vaddi, Ming-Huang Huang, Hoon Kim, Robert G. Manley

Corning Research & Development Corporation
21 Lynn Morse Drive, Painted Post, NY 14870, USA

This work studies the effect of glass substrate thermal properties on the crystallization behavior and surface topography of polycrystalline silicon (p-Si) films formed by a XeCl excimer laser annealing of amorphous silicon. 60 nm hydrogenated amorphous silicon (a-Si:H) film was deposited on different glass substrates of varying thermal properties. A 300 nm SiO_2 buffer layer was in between the glass and the silicon film. The silicon was crystallized by a XeCl excimer laser. The film properties of p-Si exhibit a small relationship with the glass substrate thermal properties. The higher thermal conductivity of glass retards the poly-Si <111> texture and reduced the residual tensile stress. The higher thermal diffusivity of the glass substrate causes a decline in the p-Si <111> texture, the surface asymmetry and flatness.

Introduction

With the increasing display area and pixel density in the thin film transistor liquid crystal display (TFT-LCD), higher mobility TFTs are required in order to maintain or shorten the charge time of pixel electrodes.[1] Amorphous silicon TFTs have a low carrier mobility which can be overcome by using polycrystalline silicon (p-Si) film as an active semiconductor layer. The large glass panels are used in display industries, which prohibit exposure to deformative high temperatures. Low-temperature polycrystalline silicon (LTPS) technology is require for high density display technologies which utilizes small transistor sizes. Excimer laser annealing (ELA) for silicon crystallization has been the preferred method for polycrystalline thin-film formation which can deliver good polycrystalline silicon films at low temperatures below 500 °C with mobility higher than 100 cm^2/Vs.[2-4] The technology has evolved over the past decades with an extensive array of literature articles and adoption in many flat panel display (FPD) manufacturing plants.[5] Much of the published research surrounding ELA has focused on mechanism and p-Si film quality improvement.[6-9] This work specifically studies the effect of glass substrate thermal properties on the silicon crystallization behavior and surface topography resulting from ELA.

Experimental details

Glass wafers, 150 mm in diameter and 0.5 mm thick were used as substrates for this study. Six different types of glass substrate were considered. The composition of the glasses is consistent to what is are used in the FPD today. They consist of a boro aluminosilicate network modified by alkaline earth species. The glasses have no intentional alkali modifiers as part of the composition. Any tramp alkali elements are <100 ppm in concentration. The glass substrate properties are shown in Table 1. A 300 nm buffer SiO_2 film was directly deposited on the substrate by plasma enhanced chemical vapor deposition (PECVD) from tetraethyl orthosilicate (TEOS) and O_2 gas mixture at 400 °C. A 60 nm hydrogenated amorphous silicon (a-Si:H) film was subsequently deposited at 400 °C in the different chamber of the same PECVD cluster system without breaking vacuum. All films were deposited by using constant deposition conditions on all glass substrates. After deposition, the sample were irradiated with a homogenized (±5%) 308 nm XeCl excimer laser with 30 ns pulse width and a fixed beam size of 4 mm × 4 mm. Samples were laser annealed with one pulse at constant energy density of 300 mJ/cm^2 in order to study the effect of glass substrate on film characteristics. The laser was rastered over a 100 mm × 100 mm region of each glass wafer.

TABLE I. Glass substrate properties.

Glass type	Density (g/cm³)	CTE (×10⁻⁷/°C)	Thermal conductivity (W/m·k)	Thermal diffusivity (cm²/sec)	Specific heat (J/(kg·K)
Glass A	2.500	36.4	Not Measured	Not Measured	1000.1
Glass B	2.385	31.7	1.513	0.0055	1155.4
Glass C	2.596	35.0	1.473	0.0058	940.8
Glass D	2.518	35.6	Not Measured	Not Measured	Not Measured
Glass E	2.520	33.0	1.303	0.0055	949.2
Glass F	2.576	37	1.409	0.0057	1001.8

Following laser annealing, surface topography or the surface roughness of p-Si was measured by an atomic force microscopy (AFM) with a measurement window of 10 μm × 10 μm, and 10 to 15 regions were measured for each sample. In-plane X-ray diffraction (IPXRD) was used to evaluate the silicon film crystal properties along with in-plane orientation of p-Si grains. Grazing incidence X-ray diffraction (GIXRD) was employed to quantify the residual stress in p-Si film. To confirm the XRD result, cross-sectional scanning transmission electron microscopy (STEM) and nanobeam beam diffraction (NBD) analysis were performed on the samples to examine the nanostructure of the resulting films on Glass C and confirm polycrystalline silicon formation base to the laser setup.

Results and discussion

The surface roughness of the p-Si film on different type of glass substrates were measured by AFM. Four roughness parameters (R_q, R_a, skewness and kurtosis) were extracted from the data, and the results are shown in Figure 1. The means diamonds are a graphical illustration of the *t*-test. The middle line of each diamond represents the group mean, and the vertical span of each diamond indicates the 95% confidence intervals of the mean of each group. Overlap marks appear as lines above and below the group mean. Overlap marks in one diamond that are closer to the mean of another diamond than that diamond's overlap marks indicate that those two groups are not statistically different at the given confidence level.[10] The variation of roughness around the same sample is less than 5%, all the samples have similar roughness level expect p-Si on Glass A. The p-Si on Glass A has larger R_q and R_a value, and smaller skewness and kurtosis, which means higher, more blunt Si hillocks formed on the p-Si film surface. Based on the calculation methods of R_q, R_a, skewness and kurtosis, as expected, R_q and skewness exhibit a clear relationship with R_a and kurtosis. The plots of R_q vs R_a, and skewness vs kurtosis are shown in Figure 2. Only R_q and skewness were selected and analyzed in following section to present the p-Si surface roughness parameters. Roughness parameters as function of glass thermal properties were plotted. Only a weak relationship between skewness and thermal diffusivity was found, as show in Figure 3.

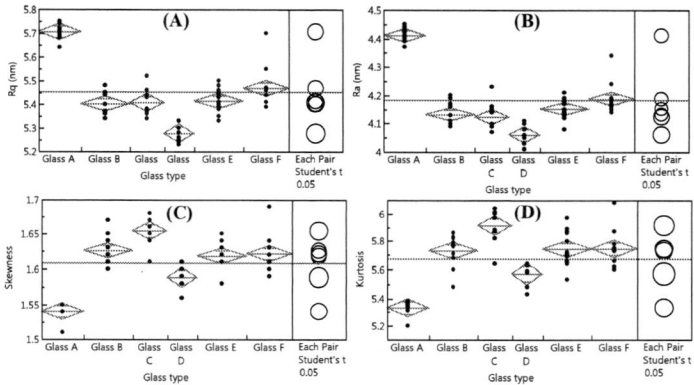

Figure 1. The p-Si film surface roughness on different type of glass substrates, including Rq (A), Ra (B), skewness (C) and kurtosis (D).

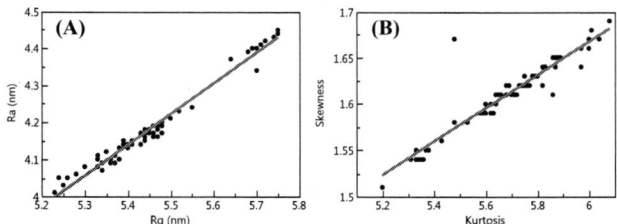

Figure 2. The relationship between Ra and Rq (A) and skewness and kurtosis

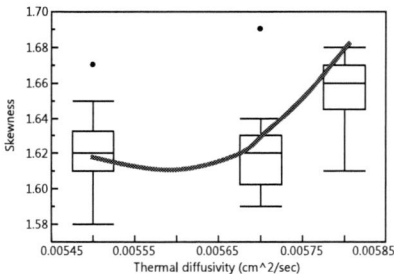

Figure 3. Skewness as a function of glass thermal diffusivity.

Due to thickness and poor crystallinity, the p-Si film resulted in a weak signal using the Bragg XRD method. In-plane X-ray diffraction (IPXRD) was used to evaluate crystalline in-plane orientation of p-Si. Figure 4 shows the IPXRD spectrum of p-Si film on glass C. The film showed (111), (220), and (311) preferred orientations plane crystalline Si samples. In order to study the orientation variation between different samples, the integrated peak area of all the orientation was calculated, and the peak area percentage of (111), (220) and (311) was obtained and listed in Table II. Also, the full width at half maximum (FWHM) of (111) peak was included in the table, which is associated with the grain size of crystals in the film. As IPXRD results cannot be used in film residual stress, so the grazing incidence X-ray diffraction (GIXRD) was employed to quantify the residual stress in p-Si film using $sin^2\psi$ method [11], the stress calculation results were catalogued in Table II. As composition and properties of studied glassed are similar, p-Si crystallinity and residual stress on all glasses resulted in no significant difference.

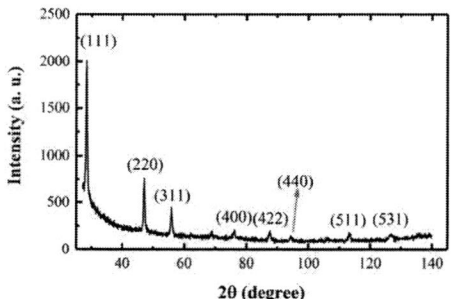

Figure 4. IPXRD spectrum of p-Si film on glass C

TABLE II. Calculation results of IPXRD and IPXRD.

Glass type	(111) peak FWHM (degree)	(111) peak area pct. (%)	(220) peak area pct. (%)	(311) peak area pct. (%)	Tensile residual stress (MPa)
Glass A	0.42	48.7	18.3	11.1	1657
Glass B	0.43	47.5	18.6	10.7	1588
Glass C	0.42	46.8	19.9	10.8	1495
Glass D	0.43	47.7	19.3	11.6	1587
Glass E	0.43	48.0	19.4	10.6	1675

Glass F	0.44	47.7	19.3	10.9	1667

Based on the correlation between p-Si surface roughness and crystallinity analysis, the (111) peak area percentage showed clear connection with R_q and skewness mean. (111) peak area percentage as a function of R_q and skewness mean were plotted in Figure 5(A) and (B), respectively. As the figures shown, better (111) orientation preferred p-Si film had rougher surface, but blunt Si hillocks formed on the surface. From the glass properties aspect, p-Si (111) orientation is affected by glass thermal conductivity and diffusivity, and tensile residual stress is influenced by thermal conductivity. From the Figure 6 and 7, the higher glass thermal conductivity and thermal diffusivity inhibit the p-Si (111) orientation, and higher thermal conductivity lowered the residual film stress.

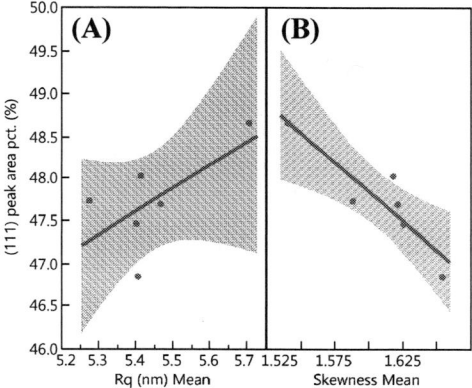

Figure 5. (111) peak area percentage of p-Si film as a function of Rq mean value (A) and skewness mean value (B) of each sample

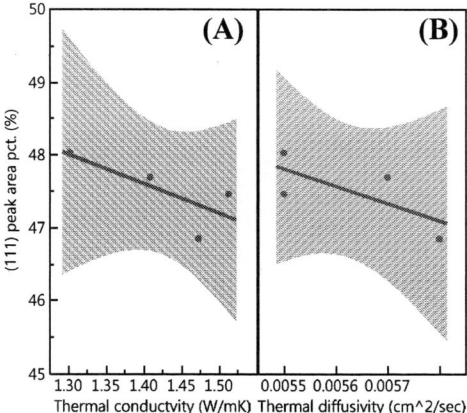

Figure 6. (111) peak area percentage of p-Si film as a function of glass thermal conductivity (A) and thermal diffusivity (B).

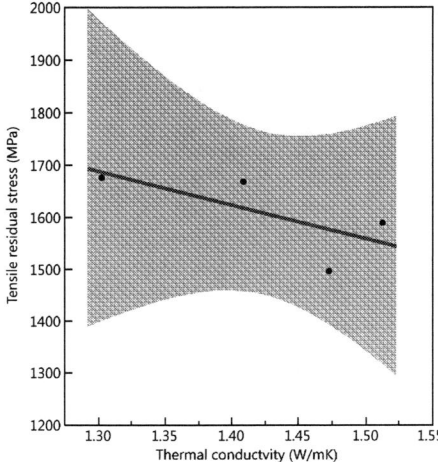

Figure 7. Residual tensile stress of p-Si film as a function of glass thermal conductivity.

STEM imaging, as well as NBD, were performed on the glass C sample in order to confirm the p-Si formation. Figure 8(A) shows the bright-field images of the laser annealed sample. The a-Si layer crystallized to find-grain p-Si, the p-Si crystalline orientation formed the contrast of image in the p-Si layer. NBD pattern approved the foundation of polycrystalline phase as shown in Figure 8(B).

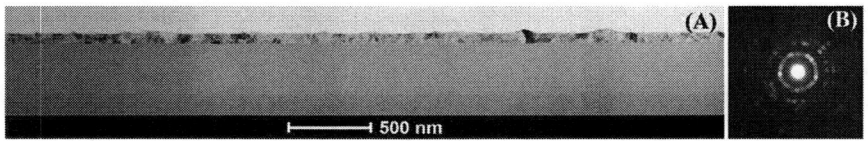

Figure 8. Cross-sectional bright-field STEM images of p-Si film on glass C (A) and NBD pattern (B).

Conclusions

The effect of glass thermal properties on the crystallization behavior and surface topography of ELA p-Si was investigated. As composition and properties of used glass substrate are similar, p-Si surface roughness, crystallinity and residual stress on all glass are similar. The p-Si surface roughness showed some connection with crystallinity behavior, higher <111> texture p-Si film had rougher surface but blunt Si hillocks on the surface. The glass thermal properties slightly affected the film properties of ELA p-Si. The higher thermal conductivity of glass retards the p-Si <111> texture and reduced the residual tensile stress, and higher thermal diffusivity declines the p-Si <111> texture and the surface asymmetry and flatness. In display industries, the light shield is introduced between glass and buffer layer in the LTPS thin-film transistors (TFTs) structure, which further weakens the effect of glass substrate on ELA p-Si properties due to the substantially higher thermal conductivity of the metal layer.

References

1. C. Yi, S. Rhee, J. Ju, S. Yim, and H. Min, *J. Mater. Sci.: Mater. Electron*, **12**, 697(2001).
2. Y. Kuo and P. Koziowski, *Appl. Phys. Lett.***69**, 1092 (1996).
3. H. Lim, B. Ryu and J. Jang, *Appl. Phys. Lett.*, **66**, 288(1995).
4. M. Furuta, T. Kawamura, T. Yoshika and Y. Miyata, *IEEE Trans. Electron Devices* **40**, 1694(1993).
5. A. Adikarri and S. Silva, *J. Appl. Phys.***97**, 114305 (2005).
6. H. Azuma, A. Takeuchi, T. Ito, H. Fukushima, T. Motohiro, and M. Yamaguchi, *Sol. Energy Mater. Sol Cells*, **74**, 289(2002).
7. K. Yamamoto, A. Nakashima, T. Suzuki, M. Yoshimi, H. Nishio, and M. Izumina, *Jph. J. Appl. Phys.*, **33**, L1751(1994)
8. W. Yeh and M. Matsumura, *Jph. J. Appl. Phys.*, **38**, L110 (1999)
9. C. Lew and M. Thompson, *MRS Online Proceedings Library Archive*, **685E**, D5.22.1 (2001)
10. M. Proust, *jmp® Version 10: Basic Analysis and Graphing, Second Edition*, p. 143, (2012).
11. S. Skrzypek, and A. Baczmanski, *JCPDS-International Centre for Diffraction Data*, **44**, 134 (2001)

Influence of Glass Surface Modification on Thin Film Copper Electrodes

Ming-Huang Huang, Ying Shi, Bin Zhu, Rajesh Vaddi, Hoon Kim, Robert G. Manley

Corning Research and Development Corporation, 1 Riverfront Plaza, Corning, New York 14830, USA

> We investigate the impact of the glass surface on the structural and electrical properties of Ti/Cu film stacks as well as single Ti adhesion layer, by modifying Corning® Eagle XG® glass surfaces using four acids common in TFT-LCD microfabrication processes. Atomic force microscopy results indicate significant change of the surface topography of Ti and Cu, particularly found on hydrofluoric acid treated glass surfaces. The film microstructures are characterized by a comprehensive analysis using Bragg, grazing incidence and in-plane X-ray diffraction, and X-ray reflectivity methods as well as the transmission electron microscopy. The sheet resistance was measured using the four-point probe method. We demonstrate the role of the glass surface state through its impact on the structural and electrical properties of Ti/Cu electrodes emanating from glass surface treatments.

Introduction

Large-area and ultra-high-resolution displays continue to progress driven by new performance requirements and applications. As display sizes become larger, the industry demands metal electrodes with low resistance to reduce the gate delay time and signal distortion. Cu is an attractive candidate due to its relative low resistivity and its superior resistance to electromigration as opposed to materials such as Al or Mo. For bottom gate TFT technologies, the metal electrode is deposited directly on the glass substrate (1). Cu is known to have poor adhesion with oxide materials such as the substrate glass. Typically, an adhesion layer such as Ti, Mo, or other Cu alloy materials (2-5) is required to improve the adhesion of Cu to the substrate. In this work, we modify Corning® Eagle XG® glass surfaces by various acid treatments to study the impact of the glass surface on the structural and electrical properties of Cu. Ti was used as the adhesion layer in this study. Atomic force microscopy is used to examine the surface morphology and the surface roughness of bare glass with different surface treatments, single Ti-adhesion layer and Cu layer of Ti/Cu film stacks. The crystal structures of both Ti and Ti/Cu film stacks are characterized by a comprehensive X-ray diffraction analysis using Bragg, grazing incidence, and in-plane configurations. X-ray reflectivity is applied to Ti film to reveal its bulk structure, such as density, thickness and surface/interface roughness. Transmission electron microscopy is also used to investigate the structure of the Ti and Cu films. The sheet resistance was measured with a four-point probe measurement. We demonstrate that the different surface morphologies of bare glass substrates derived by different treatment are preserved in the following deposited Ti/Cu thin films with critical effects on their structural and electrical properties.

Experimental

Surface modification was conducted on 0.5 mm thick Corning® Eagle XG® glass substrates with various wet chemistries commonly used in TFT-LCD fabrication processes. An example of acid surface modification is listed in Table 1. After surface treatment, all glass substrates were rinsed with deionized water with a resistivity of >18 MΩ and followed by Nitrogen blow dry. They were then loaded into a load-locked, cluster sputtering tool to deposit Ti and Cu. Prior to film deposition, the glass wafer was pre-heated in a degas chamber under low vacuum. A 20 nm Ti film was directly deposited on the substrate at 50 °C as an adhesion layer and 350 nm Cu films were subsequently deposited at 160 °C in the different chamber of the same sputter system without breaking vacuum. The deposition chamber was pre-heated to desire temperature and stabilized over one hour before deposition. The thermal anneal of Ti/Cu film was conducted in a plasma-enhanced chemical vapor deposition chamber at 350 °C for 6 minutes, but without turning on the reactant gases to simulate the typical thermal cycle of silicon nitride and oxide gate insulator films. The sheet resistance (R_s) was measured by four-point probe method with a 49-point mapping scheme to calculate the average and standard deviation. Surface roughness was characterized by atomic force microscopy (AFM) on the Veeco diDimension Icon instrument. Measurements were made in 2 μm x 2 μm scan area with 256 x 256 pixels. Three measurements were performed on each glass substrate.

TABLE I. An example of surface modification on glass substrates

Condition	Chemistry	Temperature (°C)	Time (min)
T1	HF	40	20
T2	H_2SO_4	40	30
T3	HCl	40	5
T4	HNO_3	75	30
Ctrl	N.A.	N.A	N.A

The microstructure of Cu film is characterized by cross-section tunneling electron microscopy (X-TEM). The perimeter, area, and number of Cu grain domain on each treated glass are processed on X-TEM images using ImageJ software. All the x-ray measurements, including x-ray diffraction (XRD) and x-ray reflectivity (XRR), are carried out using PANalytical Empyrean x-ray diffractometer equipped a Cu tube with power setting of 45 kV and 40 mA. All the XRD patterns are analyzed using the software package HighScore Plus. Three configurations are used to collect XRD patterns of thin film samples for a comprehensive characterization: Bragg, graze incident (GI) and in-plane (IP) XRD. Both Bragg and GI scans are collected using the parallel x-ray beam optic Bragg-Brentano HD (BBHD) with line-shape beam, coupled with a 1/4 divergence slit (DS), 1/8 incident beam anti-scatter slit and a 20mm mask. The GI incident angle for both Ti-single layer and Ti/Cu-bilayer is 0.5°. The IP scan is collected using the x-ray lens optic with point-shape beam, coupled with divergence slit fully open at 10 mm and a narrow 0.5 mm mask. The sample tilt angle is set up at 89.5° for Ti-single layer coating which corresponding to 1° incident angle, and 89° for Ti/Cu-bilayer coating which corresponding to 2° incident angle. The x-ray reflectivity (XRR) measurement uses the parallel x-ray beam optic Bragg-BrentanoHD (BBHD) with line-shape beam, coupled with a 1/16 divergence slit (DS), 1/32 incident beam anti-scatter slit and a 0.1 mm Cu beam attenuator. The beam divergence from this setup is narrow enough to generate clear XRR fringes for the 20 nm Ti-single layer, but it is too broad for the 20 nm Ti / 350 nm Cu bilayer film stack, which needs high-resolution beam monochromator to produce the distinguishable fringes for data modeling. The XRR pattern is analyzed using X'Pert Reflectivity software package.

Results and Discussion

We first characterize the surface roughness and morphology of as-treated glass surface, deposited with 25nm Ti only, and 25 nm Ti/ 350 nm Cu film stack. Figure 1 shows surface roughness of treated glass with T2, T3, and T4 conditions slightly increases as compared to the control condition. The corresponding Ra and Rq values increase from 0.2 to 0.3-0.4 nm. Conversely, the T1 treated glass surface becomes much rougher with the Ra and Rq values increasing to ~1.25 nm. This could be attributed to a certain degree of glass etching with HF containing chemistry as compared to glass leaching with other acids. The impact of glass surface roughness is also reflected on 25 nm Ti and 350 nm Cu surfaces, which it becomes more pronounced on Cu film. The Ra and Rq values of Cu on T1-treated glass increases to >3 nm and the variation also increases as Cu film is much thicker. The selected surface morphologies on control and T1 conditions are shown in Figure 2. The AFM images of T2-T4 conditions are not shown here as the results are not much different from control condition. There are tiny white spots shown in Figure 2(d), which may be correlated to the precipitates left on the surface after T1 treatment. The surface morphologies for Ti and Cu on T1-treated glass are noticeably different from the control and T1 conditions.

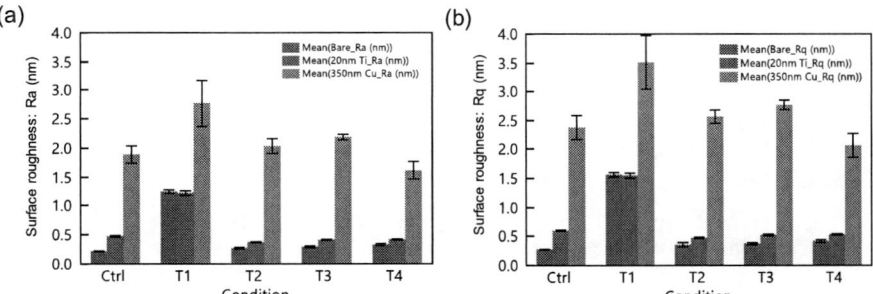

Figure 1. Surface roughness of as-modified glass surface and Ti/Cu deposited surface for various surface treated conditions, (a) R_a values and (b) R_q values.

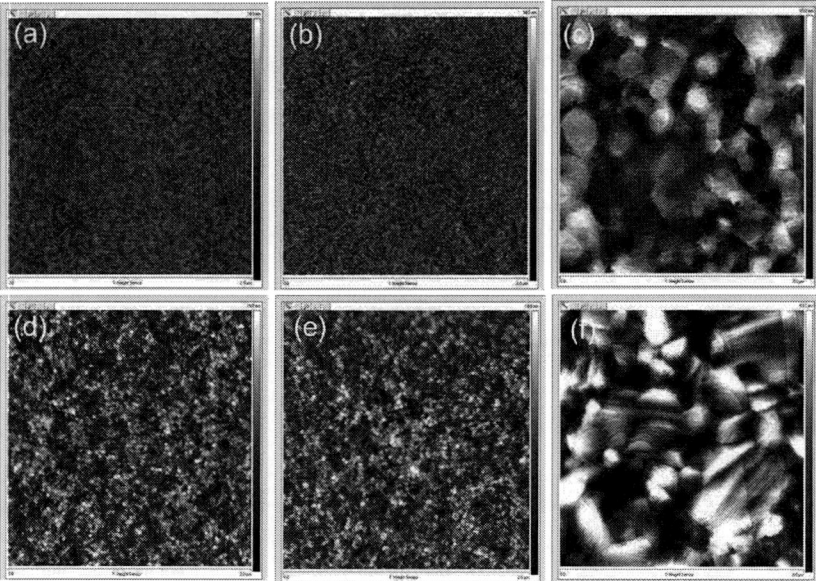

Figure 2. AFM surface morphologies on control (top row) and T1(bottom row) samples for bare glass surface (a) and (d), with 20 nm Ti (b) and (e), and with 350 nm Cu (c) and (f). The color scale is 0-10 nm for all images.

We measure sheet resistance of Ti and Cu films to study the impact of glass surface modification as well as the impact of thermal anneal on the electrical properties. Figure 3 shows the sheet resistance of 20 nm Ti and 350 nm Cu before and after thermal anneal. Figure 3(a) shows averaged sheet resistance (R_s) of 350 nm Cu deposited on glass treated with conditions T2 to T4 is 71-72 mΩ/$\square$, similar to the untreated control glass of 71 mΩ/$\square$. On the contrary, the averaged R_s of 350 nm Cu deposited glass treated with conditions T1 has a slightly higher value of 74 mΩ/$\square$. After thermal annealing at 350 °C for 6 minutes in vacuum, the R_s values of all conditions are reduced to 65 mΩ/$\square$, except T1 condition is slightly higher (67 mΩ/$\square$). The electrical resistivity of Cu can be calculated by sheet resistance times film thickness. this small change in electrical resistivity due to the glass surface roughness may be ignored, no matter it is with and without thermal anneal. Since Cu is deposited on the top of Ti, not directly on the glass surface, one would expect that the surface roughness may have an impact on 25 nm Ti. We compare the sheet resistance within same group, *i.e.* samples having the same coating layer and thermal treatment history. As shown in Figure 3(b), the averaged R_s of 20 nm Ti deposited on glass treated with conditions T2-T4 is 105-109 mΩ/$\square$, similar to the untreated control glass is of ~107 mΩ/$\square$. On the other hand, the T1 sample shows noticeable higher value than the other four samples. Similarly, after thermal annealing at 350 °C for 6 minutes in vacuum, the R_s values of all conditions are reduced to 53-56 mΩ/$\square$, except T1 condition is slightly higher (59 mΩ/$\square$). As mentioned above, the T1 sample surface is rougher demonstrated in the AFM analysis, indicating the surface roughness has a correlation with sheet resistance.

We also notice that the Ti-single layer sheet resistances of the four surface treated samples show the same relative drop ratio with an average value of 48.8 $\pm$0.8%, as calculated by ($\Omega_{as\text{-}coated}$-$\Omega_{annealed}$)/$\Omega_{as\text{-}coated}$$\times$100%. The same relative sheet resistance drop ratio with an average value of 9.3 $\pm$0.6% is also observed for the Ti/Cu-bilayers coated on all five different surface treated substrates. Because the drop ratio is equivalent for the samples with same coating but different surface treatments, the impact that annealing effects have on sheet resistance is independent of surface roughness.

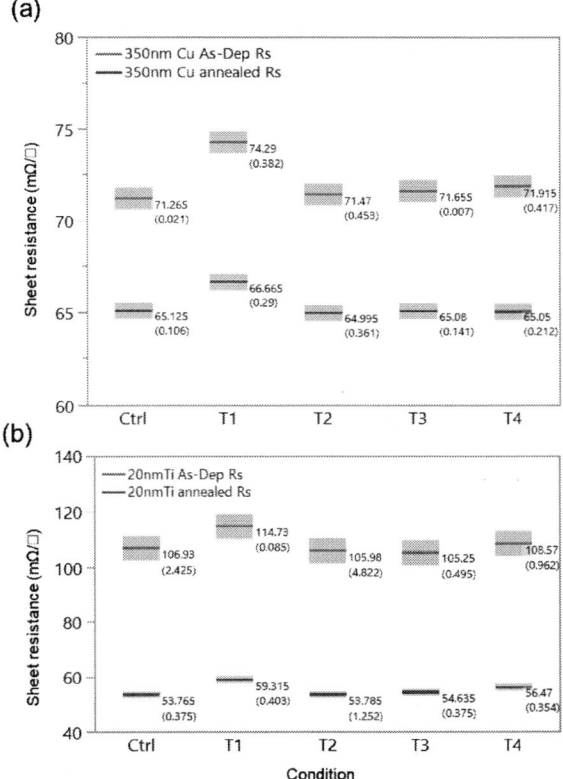

Figure 3. Sheet resistance of 350 nm Cu film (a) and 20 nm Ti film (b) deposited on different surface treated substrates. The red symbols correspond to as-deposited samples while blue symbols correspond to annealed samples.

We investigate the nano crystal structure change by XRD and interface properties between the Ti film and treated glass surface to further understand the influence of glass surface roughness as well as the annealing effort. Three sets (Bragg, GI and IP) of the XRD patterns are plotted in Figure 4 for the Ti-single layer samples coated on the surfaces treated by control (a) and T4 (b) methods. In each sub-plot, the as-coated (red) and annealed (blue) patterns are overlaid for same measurement configuration. The Ti film has a hexagonal close-packed (HCP) structure and grows along (002) direction with a strong preferred orientation, supported by the only (00l) peaks detected in the Bragg scans. In contrast, IP-scans only show (hk0) peaks which are the planes parallel to the growth direction. With GI scan showing the peaks in between, it clearly demonstrates how the three configurations provide a whole view of the thin film sample. There are no observable differences between two surface treated samples, however both samples show significant peak shifting after annealing. Comparing the blue pattern (annealed) with red ones (as-coated) in each overlay, all the peaks in all three configurations shift to lower 2θ angle, indicating a larger lattice constant. The lattice constants (LC) can be quantified from three XRD patterns of each sample. The LC-c can be derived from (002) peak of Bragg scan, while LC-a can be derived from (hk0) peak of IP scan and both LC-a and c can be obtained from GI scan. Crystallite size cannot be calculated reliably from the only one (002) peak in Bragg scan, instead the full width at half maximum (FWHM) is listed in Table II, to compare the change by annealing. Both LC-a and c increase after annealing with LC-c increasing more than LC-a. This leads to a higher c/a ratio, towards to the ideal c/a ratio of 1.633 for HCP phase, while the bulk Ti HCP phase has a low c/a ratio of 1.583 [6]. We conclude annealing leads to higher c/a ratio. The FWHM of (002) peaks get narrower for both samples, indicating the crystals grow larger in the direction parallel to growth direction.

TABLE II. Annealing effect on lattice constants and crystallite size/strain of Ti-single layer

Condition	Thermal anneal	Lattice constants (Å)						FWHM of (002)
		LC-c		LC-a		c/a		
		GI	Bragg	GI	IP	Bragg/IP	GI/GI	Bragg
Ctrl.	Yes	4.760	4.775	2.939	2.946	1.625	1.620	1.01
	No	4.699	4.725	2.947	2.938	1.603	1.594	1.08
T4	Yes	4.819	4.819	2.963	2.970	1.626	1.626	1.07
	No	4.727	4.725	2.927	2.940	1.614	1.615	1.10

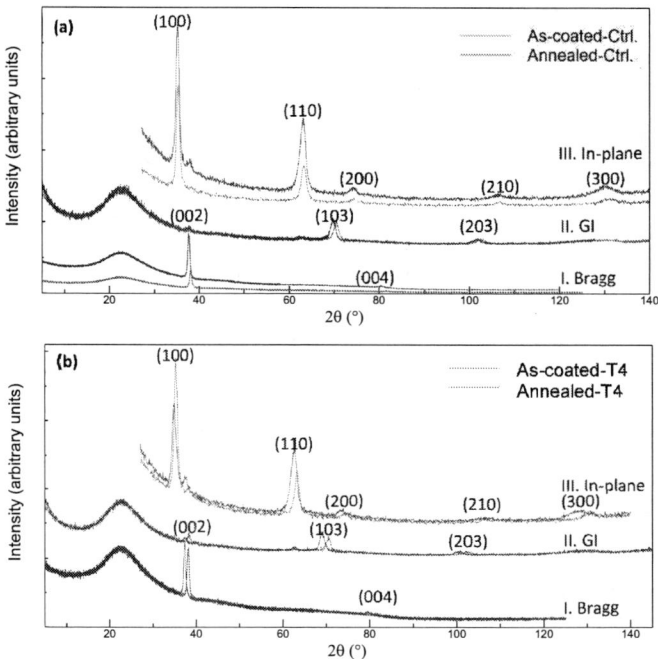

Figure 4. Annealing effect on X-ray diffraction patterns of the Ti-single layer coated on control (a) and T4 treated (b) glass surfaces. For each surface treatment coating, as-coated (red) and annealed (blue) samples are measured by three configurations, Bragg (I), Graze-incident (II) and In-plane (III). The Bragg scans are scaled down to the similar intensity level of GI and IP scans, the GI and IP scans are moved up for clarity.

Due to the instrumental limitation, XRR measurements are only done for the Ti-single layer. The as-coated (red) and annealed (blue) XRR patterns for Ti-single layer coated on glass substrate treated by control and T4 methods are plotted in Figure 5 (a) and (b), respectively. Annealing changes XRR patterns in two significant ways: 1) comparing to as-coated XRR pattern, annealing pattern shows higher critical angle ω_c, indicating higher packing density; 2) the fringes of annealing XRR decay faster than those of as-coated one, indicating the rough Ti-glass interface. The X'Pert Reflectivity software is used to derive information, such as density, thickness and interface/surface roughness. A thin TiO_2 layer (~ 4 nm) is added as the top surface for a better model fitting, which is not surprising since Ti metal is easily oxidized in air. Only the Ti-related structure information are plotted in Figure 6, which including density, thickness and Ti-glass substrate interface roughness. Annealing increases both density and interface roughness. We deduce that the density of Ti-layer increases after thermal anneal, i.e., thermal densification, is the primary factor for sheet resistance drop, where the interface roughness is a secondary factor.

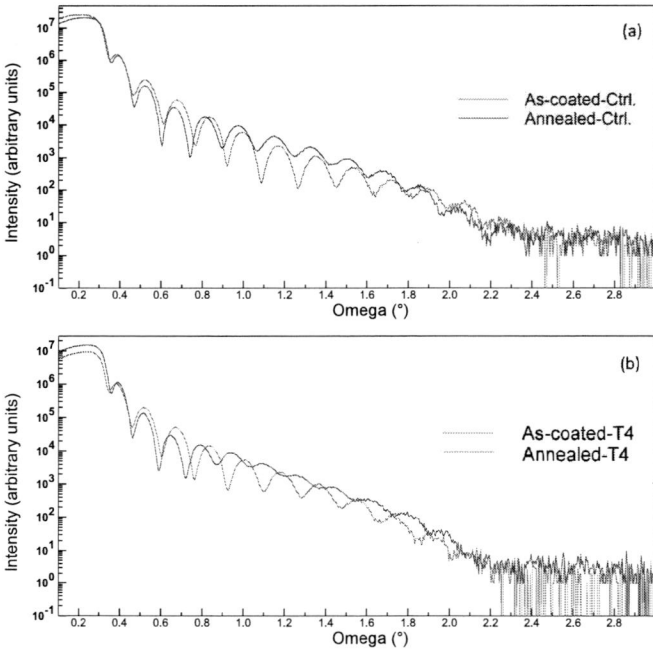

Figure 5. Annealing effect on x-ray reflectivity patterns of Ti-single layer coated on control (a) and T4-treated (b) glass surface. As-coated (red) and annealed (blue) XRR are overlaid for each surface treatment coating

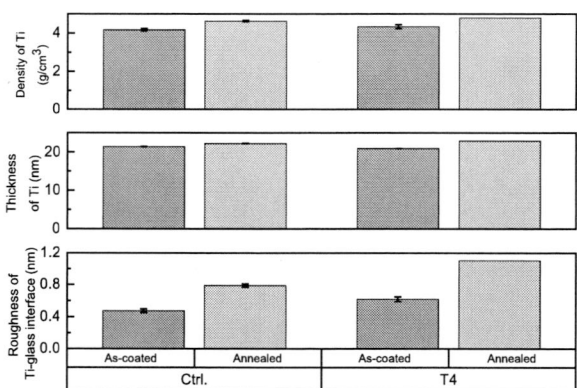

Figure 6. Structural information (density, thickness and Ti-substrate interface roughness) derived from XRR for as-coated (red) and annealed Ti-single layer coated on Ctrl. and T4 treated surface. The error bar is calculated from three independent XRR measurements.

Despite the effect of annealing dominating Ti and Cu film microstructures, the impact of glass surface roughness still plays a role on the film structures within the same thermal treatment. Three sets (Bragg, GI and IP) of the XRD patterns are plotted in Figure 7 for the annealed Ti-single layer (a) and Ti/Cu-bilayer (b) coated on the surfaces treated by T1 (red) and Ctrl. (blue) methods. As mentioned in the annealing effect study, the Ti phase of Ctrl. sample grows with a strong (002) preferred orientation, while T1 sample shows a much weaker (002) preferred orientation with broad peak shape and much lower intensity. The same peak shape and intensity difference is also observed for IP scans of Ctrl. and T1 samples, the extra (101) peak observed in the IP scan of T1 sample adding the additional support for its weak preferred orientation. Only the cubic-Cu phase is identified in the Ti/Cu-bilayer samples as shown in Figure 7(b). Compared to the HCP-Ti phase, all the peaks in Cu-phase are much sharper because the much thicker Cu layer, 350 nm, compared to Ti layer of 20 nm. Similar to the Ti-single layer sample, the control sample has a Cu-phase with strong (111) preferred orientation as shown in its Bragg scan, only one (220) peak is observed in its GI and IP scans, while the T1 sample shows many more peaks, indicating a much weaker preferred orientation.

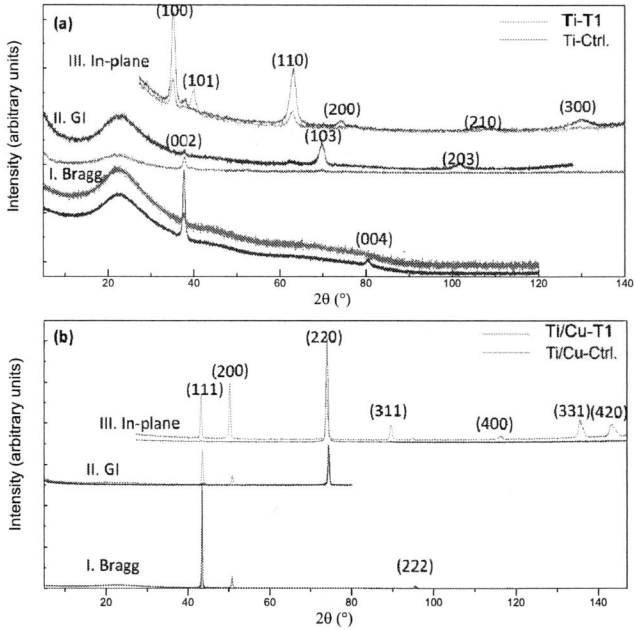

Figure 7 Surface treatment effects on X-ray diffraction patterns of Ti-single layer (a) and Ti/Cu bilayer coatings (b). For each coating structure, T1 (red) and control (blue) surface treated samples are measured by three configurations, Bragg (I), Grazing-incidence (II) and In-plane (III). The Bragg scans are scaled down to the similar intensity level of GI and IP scans, the GI and IP scans are moved up for clarity.

The preferred orientation in Cu film is also confirmed in X-TEM analysis as shown in Figure 8. A preferred orientation of annealed 350 nm Cu film is observed on T2-T4 samples, while less preferred orientation of Cu film is shown in T1 samples. X-TEM images are further processed to analyze the grain domain counts, perimeter, and area for each treat condition. For a given image region, T1 sample has a higher domain count and smaller domain profile perimeter and area as compared to the rest of three samples.

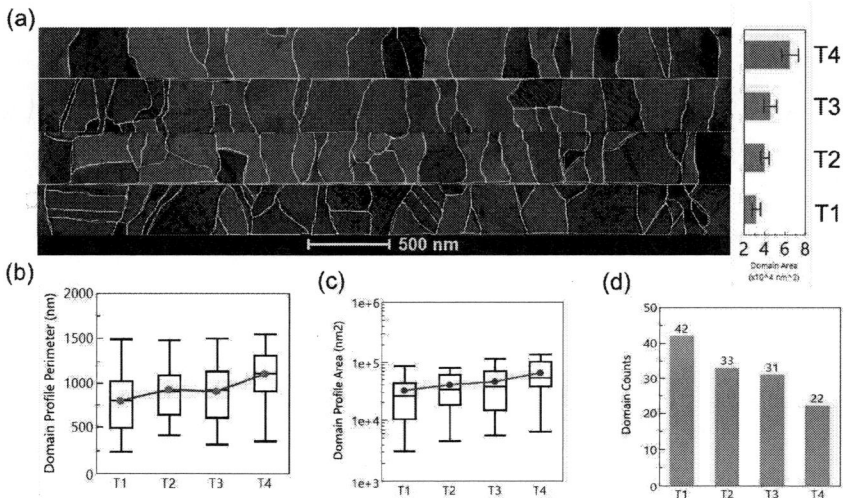

Figure 8. X-TEM image analyses for annealed 350nm Cu film on modified glass substrate. (a) X-TEM images with highlighted grain domains and estimated domain area; (b) Estimated domain profile perimeter; (c) domain profile area; (d) domain counts for a given image region.

The XRR patterns for annealed Ti-single layer coated on glass substrate treated by four different methods are plotted in Figure 9. Significant difference of the fringe-decaying can be observed among all patterns. The same structural information is derived from XRR using X'Pert Reflectivity software with the Ti-related structure information plotted in Figure 10, which including density, thickness and Ti-glass substrate interface roughness. As shown in Figure 11, a linear correlation is derived between sheet resistance and roughness of Ti/glass interface with $R_2 = 0.96$. This demonstrates that for the samples with same thermal history, the interface roughness is the determining factor for sheet resistance. From the reliability point of view, (111) textured Cu film has advantage as well.

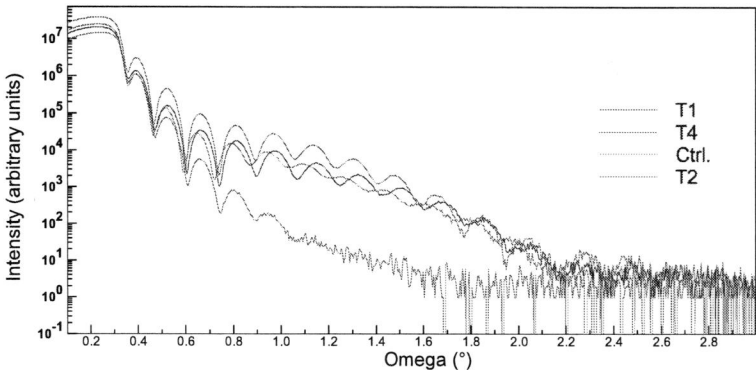

Figure 9. Surface treatments effect on x-ray reflectivity patterns of Ti-single layer coated on glass surfaces treated by four different methods.

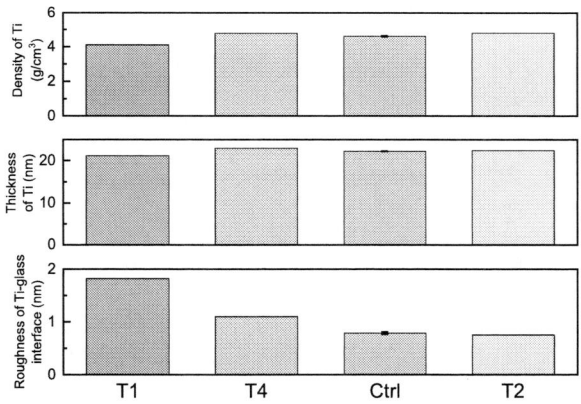

Figure 10. Structural information (density, thickness and Ti-substrate interface roughness) derived from XRR for as- annealed Ti-single layer coated on glass substrates with four different surface treatments. The error bar is calculated from three independent XRR measurements.

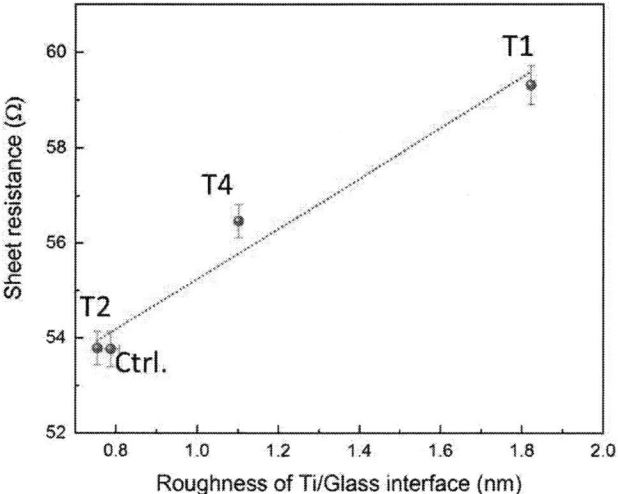

Figure 11. Relationship between sheet resistance and roughness of Ti/Glass interface. A linear fitting is obtained as shown by dotted line with R^2=0.96.

Conclusions

We investigate the impact of the glass surface on the structural and electrical properties of Ti/Cu, single Ti adhesion layer, by modifying Corning® Eagle XG® glass surfaces using four common acids used in TFT-LCD fabrication processes. We compare the structural and electrical properties of Ti/Cu films deposited on treated glass surfaces and demonstrate a linear relationship between sheet resistance of Ti and roughness of the Ti/Glass interface. We conclude annealing leads to a higher c/a ratio for Ti layer. The narrower FWHMs of (002) XRD peak indicates the crystals grow larger in the direction parallel to growth direction. Annealing increases both density and interface roughness. We conclude that the density of Ti-layer increases after thermal anneal is the primary factor for sheet resistance drop, where the interface roughness is a secondary factor. With the same thermal history, the influence of surface roughness was observed on the preferred orientation of annealed 350 nm Cu film. X-TEM image analysis indicates the higher surface roughness leads to a higher domain count and smaller domain profile perimeter and area. Minimizing the roughness of glass surface during surface cleaning is critical for low resistance Cu film formation. As we have learned glass surface treatment can alter Cu film microstructures. Future work includes to further investigate how the etching rate responds to the change of Cu film properties.

Acknowledgments

The authors would like to thank Minghui Zhang for X-TEM characterization and Alexis Tindall for AFM measurements.

References

1. HIS market technology, "Cu electrode, the key to UHD LCD technology", August 13, (2014)
2. J. Koike, K. H., M. Naito, P. Yun, and Y. Sutou, "P-33: Cu-Mn Electrodes for a-Si TFT and Its Electrical Characteristics" in Proc. SID 2010, p. 1343 (2010)
3. W. H. Lee, H. L. Cho, B. S. Cho, J. Y. Kim, W. J. Nam, Y-S. Kim, W. G. Jung, H. Kwon, J. H. Lee, and J. G. Lee, *Appl. Phys. Lett.* 77, 2192 (2000)
4. W. Lee, H. Cho, B. Cho, J. Kim, Y. Kim, W.Jung, H. Kwon, and J. Lee, *Journal of Vacuum Science & Technology* A 18, 2972 (2000)
5. H. Sirringhaus, S. D. Theiss, A. Kahn, and S. Wagner, *IEEE Electron Device Lett.* 0741-3106, 18, 388 (1997)
6. Yang, J.X., Zhao, H.L., Gong, H.R., Song, M. & Ren, Q.Q., "Proposed mechanism of HCP → FCC phase transition in titanium through first principles calculation and experiments," Sci. Rep., vol. 8, p. 1992 (2018)

Edge-oriented LTPS via Flash Lamp Annealing Using a Cr Adhesion Layer for Improved Wettability

G. Packard [a], A. Rosenfeld [a], M. Hum [b], R.G. Manley [c], and K.D. Hirschman [a]

[a] Department of Electrical and Microelectronic Engineering, Rochester Institute of Technology, Rochester, New York, 14623, USA
[b] Department of Materials Science, Rochester Institute of Technology, Rochester, New York, 14623, USA
[c] Corning Incorporated, Science & Technology Research, Corning, New York, 14831, USA

> PMOS TFTs built on flash lamp annealed polycrystalline silicon are presented. A thin layer of chromium is used as an adhesion promoter to prevent randomized voids from forming, allowing predictable grain growth in an elongated morphology oriented away from silicon mesa edges. Operational TFTs were realized, revealing that the Cr under-layer does not result in a significant conductance in parallel with the channel, and effectively serves as a barrier to glass contaminants. The methodology supports the production of TFTs with a high degree of electrical uniformity. Comparisons are made between electrical characteristics of transistors with grains oriented in the same direction as the channel carrier pathway versus those with perpendicular orientation; the former demonstrating approximately ten percent higher channel mobility. Process details and representative electrical characteristics are described.

Introduction

The flat panel display (FPD) industry is increasingly in search of methods to improve the fabrication process of thin film transistors (TFTs). As the market for displays of all sizes and applications increases, the need for additional technologies to satisfy these niches becomes more and more relevant. Such processes and materials must be uniform, replicable, able to achieve high carrier mobility and low leakage currents, and above all easily scalable to large-area panels at low cost. The limitations of glass and low-temperature substrates renders this field a separate challenge than that of traditional ULSIC scaling (1).

The industry leader for high quality FPD TFTs remains low-temperature polycrystalline silicon (LTPS) crystallized through excimer laser annealing (ELA). Here, amorphous silicon is deposited on a glass substrate at low temperature and crystallized in-situ with the exposure of a charged excimer laser (usually XeCl) with an emission spectrum in the UV range. Though this method thoroughly satisfies many of the requirements listed above, it suffers in considerations of cost and scalability. ELA requires repeated laser pulses of a small and non-continuous laser spot, which must be scanned across the entire area to be crystallized. As the display industry trends towards larger, Gen 10+ substrates, the cost and logistics of this method become increasingly demanding.

Other methods are rapidly reaching technological maturity as well, such as annealing amorphous silicon with a continuous-wave laser in the blue light regime rather than an excimer that must be repeatedly charged and discharged (2). Though it requires a vastly reduced power budget, this process still involves a raster scan of the full area that must be crystallized. In a non-silicon strategy, amorphous oxide semiconductors are being used to cheaply produce uniform NFETs with very low leakage current. Recent progress has been made in integrating a blue-laser annealed silicon PFET with NFETS built on IGZO, allowing the strengths of each technology to mitigate the limitations of the other (3).

Flash lamp annealing (FLA) has been considered an alternative or co-current strategy to ELA for LTPS fabrication in some applications. Rather than using the precision of an exciplex UV laser, FLA exposes large areas of amorphous silicon to a single, high-intensity pulse from a xenon flash lamp. The emission spectrum of Xe overlaps sufficiently with the absorption of a-Si to allow selective energy transfer leading to a melting and recrystallization of silicon, all without the need for multiple laser sweeps or targeted precision. The main advantages of FLA are thus in the same areas in which ELA most lacks: a single FLA exposure can crystallize an area of silicon on the order of 100 cm^2 and is theoretically scalable to any size substrate with the addition of careful masking and a scanning multi-shot apparatus (4).

Preliminary Work

Previous studies into FLA LTPS TFTs have been plagued by issues with material uniformity (5). The low wettability of liquid silicon on most low-absorption materials has often resulted in a crystallized material made of randomized voids and pathways, as shown in Figure 1a, with the density of voids proportional to the amount of energy absorbed during the anneal. This material has still been used to produce functional and high-performance NMOS and PMOS transistors (6) in a variety of structures, such as the PMOS FLA LTPS transistor in Figure 1b with hole mobility of 44 cm^2/(Vs) and threshold voltage of -2.8 V (7). However, the operational consistency and thus ultimate device yield of these final devices has doubtless suffered thanks to this random morphology. This heavily voided LTPS structure is likely unsuitable for industrial integration, especially for processes that require very small devices.

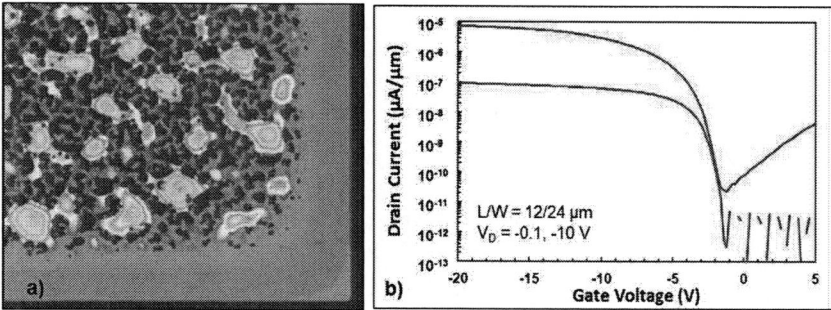

Figure 1. a) Close-up (1000x) of randomly dewetted FLA LTPS, indicating characteristic voids and contiguous pathways. Concentric "puddles" are due to optical effects of a capping oxide layer. b) Transfer characteristics of a L12W24 PMOS TFT built on such material, using a pre-amorphization implant strategy to boost boron activation. $V_T = -2.8$ V, $\mu_{chan:max} = 44$ cm^2/(Vs). (7)

Many different materials have been used to promote the adhesion of liquid silicon. There is, however, a tradeoff unique to this crystallization method regarding the addition of layers in the pre-FLA stack. As the xenon pulse is a broad-spectrum emission source, most of the energy emitted is not absorbed in the active layer and instead passes through the film. This means that any surrounding layers must be either highly transparent or thin enough that absorption is kept to a minimum. Figure 2 indicates the explosive result of an FLA process with a 10 nm layer of buried alumina (an optically transparent film at visible wavelengths), as well as a patterned molybdenum bottom gate structure that provides additional absorption area resulting in a heavily localized increase in effective energy delivery. Therefore, the addition of any new material is a careful balancing act to avoid the conditions of "too thin to be continuous or effective" and "too thick to permit the flash intensities needed to melt amorphous silicon".

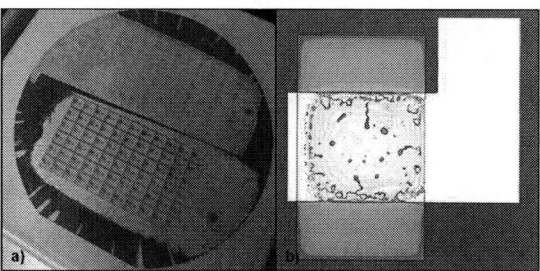

Figure 1. a) Image of the obvious and detrimental effects of a 10 nm Al$_2$O$_3$ adhesion layer after FLA. The clear/white areas indicate massive and large scale ablation. b) Micrograph of a silicon mesa crystallized with FLA while above an isolated molybdenum gate structure. The silicon above the gate structure is heavily dewetted and discontinuous, while the upper and lower sections of the mesa did not absorb enough energy to attain melt phase.

In these studies, a very thin layer of chromium was used as an adhesion layer. Previously studied metal underlayers demonstrated problems with flash lamp absorptivity, leading to large-area ablation; however, this was avoided by keeping the chromium layer as thin as was feasible and depositing it directly on the surface of the glass substrate to minimize internal reflection. In this way, the Cr underlayer serves a dual purpose of promoting adhesion and acting as a barrier between the silicon and glass layer; a remarkable and beneficial feature, as liquid silicon in direct contact with borosilicate glass will readily absorb large quantities of p-type impurities, rendering the entire layer degenerately doped and denying the possibility of a transistor off-state. The inclusion of very thin chromium allows silicon to recrystallize without forming randomized voids, leading to an edge-directed morphology with visible elongated crystal grains.

Methods

In this experiment, a 6 nm layer of chromium was deposited on 150 mm Corning Lotus display glass wafers via electron beam evaporation to promote adhesion, followed by a 60 nm layer of amorphous silicon via PECVD. The silicon was dehydrogenated at 450 °C and patterned with SF_6 RIE into large "super-mesas" to control the growth direction of the eventual crystallization. A 100 nm layer of PECVD SiO_2 was deposited as an antireflective layer. The wafers were then crystallized with a single FLA 250 µs pulse ranging between $4.8 - 5.2$ J/cm^2 (measured by blackbody bolometer; the amount absorbed by the silicon layer is approximately 30% of this) under 400 °C substrate heating, which allows for a full silicon melt and recrystallization into LTPS. The resulting material exhibited characteristic X-shaped ridges and visible grains oriented perpendicular to the edges of the super-mesas.

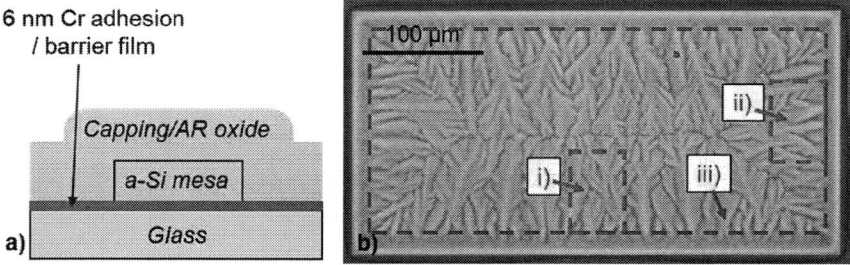

Figure 2. a) Diagram of wafer stack prior to FLA (not to scale). b) Micrograph of a polycrystalline silicon mesa crystallized with Cr-adhesion enhancement, demonstrating characteristic edge-directed grain morphology. Labels i), ii), and iii) indicate hypothetical ways in which this mesa could be reduced to provide a WG channel, CG channel, or full-mesa device.

The super-mesas were then patterned into final device mesas in three separate varieties as shown in Figure 3b: i) those with channel pathways oriented in the same direction as the visible grains: with-grain (WG), ii) those with channel paths oriented perpendicular to visible grains: cross-grain (CG), and iii) those that largely encompassed the entire super-mesa body. Source and drain regions were formed on each mesa by high-dose boron ion implant, followed by a 12-hour furnace anneal at 630 °C for dopant activation. Devices were metallized with aluminum and sintered in forming gas to promote contact formation and passivate dangling silicon bonds.

Results

Material

The predictability of this material is vastly superior to FLA LTPS produced without adhesion layers. For mesas that attain this morphology, crystal grains always form as elongated domains with their major axis oriented perpendicular to the edges of amorphous silicon mesas. This direction can thus be predictably controlled by tailoring the size and shape of the super-mesas during FLA and selecting smaller regions to use afterwards based on the resultant pattern of ridges. Material similar to this has been previously obtained in crystallization processes that use ELA or CW annealing (8), (9), but the ability to produce it on a large scale at once is a significant step towards controlling the inherent variability of FLA.

Atomic force microscopy was performed on one of the edge-directed LTPS mesas, as shown in Figure 4a. In this structure, a maximum height differential of around 55 nm between the tops of ridges and the lowest point on the mesa is demonstrated (ignoring scanning artifacts). This is significant as the original thickness of amorphous silicon as deposited was only 60 nm, suggesting large-scale mass transfer during crystallization. However, dewetting behavior is minimized. The elongated structure, lack of voids, and edge-inward pattern of this morphology is hypothetically consistent with the formation mechanism of explosive crystallization as opposed to quasi-equilibrium melt-phase, which results in the randomly irregular material demonstrated previously. In explosive crystallization, the energy emitted by the solidifying of melted edge regions is sufficient to cause melting of adjacent, previously non-melted amorphous silicon. This additional latent heat of fusion may provide enough energy to liquify silicon in the amorphous state as it has a lower melting temperature than the crystalline form. In this way, a crystallization wave may propagate from edges of mesas towards their center, leaving elongated crystal grains and raised ridge-like patterns where advancing fronts intersect.

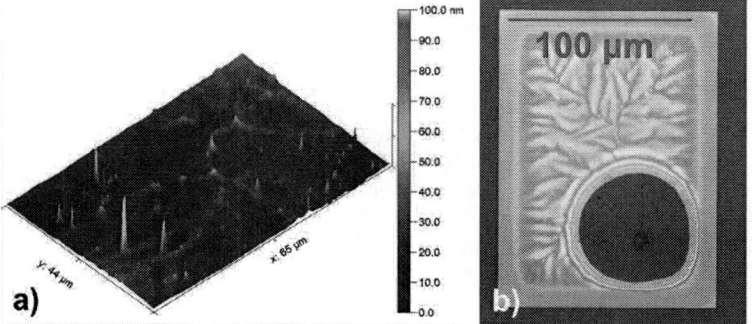

Figure 3. a) Atomic force microscopy image of a crystallized mesa indicating the relative height of ridges and boundaries. b) Micrograph of a crystallized mesa with a large circular void present. The red dashed lines are of equal length, suggesting that this void propagates outwards from a speck defect in the center and is slowed by the less-crystallized mesa border.

The uniformity and integrity of this material was not perfect, as seen in Figure 4b. After FLA crystallization, large round voids could be occasionally seen in some LTPS mesas. The sparsity and circular shape of these voids suggests that they are formed by a different mechanism than the thoroughly randomized void-like material obtainable without an adhesion underlayer. They may instead be the result of small point defects: perhaps areas where the chromium layer was broken by a pinprick flaw, or where a surface contaminant or small particle was present on the sample during a-Si deposition or FLA crystallization. The formation of these circular voids can likely be mitigated with additional control or more careful deposition methods.

Device Operation

The devices produced in this experiment were PFETs doped via ion implantation and subsequent furnace activation at a glass-compatible temperature; an option that was chosen for simplicity. Several other methods have been explored for a more complete dopant activation in FLA LTPS, such as co-implantation of electrically inert Si-ions for amorphization or secondary FLA treatments for activation. Additionally, this process is compatible with ion shower or in-situ doping.

Three forms of PFETs were produced in a variety of device dimensions: type i) WG channels, type ii) CG channels, and type iii) channels as entire super-mesa, as referenced in Figure 3 earlier. Of these, type iii) is the simplest to process, requiring no silicon etching after the crystallization step. On the other hand, this structure results in multiple directions of grain orientation present in the channel pathway. Figure 5d shows an overlay of transistor regions on one such crystallized mesa, where it can be seen that the channel pathway contains both parallel and perpendicular large grains as well as a long intersecting ridge. A sample device of this full-mesa variety with length/width dimensions of 96/96 μm is shown in Figure 5a. The device here has a threshold voltage near -7 volts, which indicates a relatively high level of interface charge. Additionally, these transistors tended to suffer from high leakage current at a drain bias of -10 V. Such issues are a common downside of LTPS devices which are exacerbated by the presence of active dopant segregation in grain boundaries; as a higher density of grain boundaries are present, the level of leakage can be expected to increase.

Figure 5b demonstrates the reproducibility of this device structure. The variability is quite low among twenty devices of the same dimensions as 5a, especially in on-state operation. The off-state leakage behavior is not as consistent, with the variation most likely associated with the mechanism responsible for the enhanced magnitude. This may ultimately be ascribed to spatial variances in flash lamp pulse delivery during crystallization. Figure 5c demonstrates the device transconductance in low drain bias operation. Using the common maximum transconductance method, a channel mobility of 35 $cm^2/(Vs)$ can be extracted.

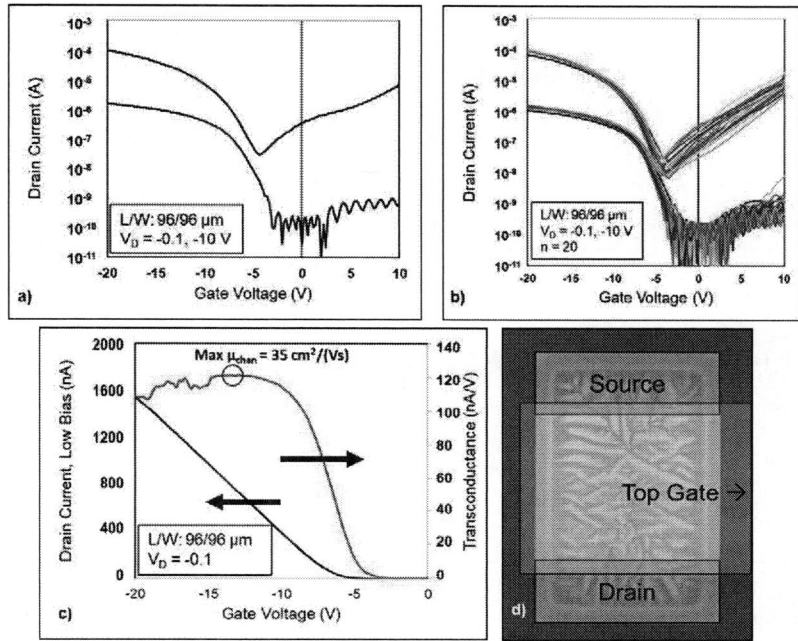

Figure 4. a) Transfer characteristics of a typical "full-mesa" style transistor of L/W = 96/96 μm. b) Overlay of transfer curves of 20 such transistors. C) Linear-scale low drain bias and transconductance of transistor in section a, indicating maximum extracted channel mobility. d) Micrograph of full-mesa style structure indicating overlay of source, drain, and channel regions.

The effect of parallel vs perpendicular grain structure was investigated by producing TFTs with both type i) WG and type ii) CG form. Figures 6a and 6b demonstrate these two schemes; of important note is the general direction of grain boundaries in between the source and drain shaded areas, as the channel is defined by the space between these wells. The CG structure is thought to contain more boundaries that interfere with the flow of current, causing scattering events that impede carrier mobility. The WG structure, on the other hand, may have single grains that bridge the entire channel.

Figures 6c and 6d demonstrate the operation of a typical pair of such devices. Again, both suffer from a high leakage current at a drain bias of -10 V, mirroring the issues of the full-mesa device above. Interestingly, the lowest current attainable was nearly an order of magnitude lower for the WG device than the CG. Additionally, the switching characteristics of CG devices appeared "lumpy" and variable, as if they were a composite of many devices that exhibit slight differences in threshold voltage and subthreshold slope. This behavior is not seen in the WG devices, which have a generally higher transconductance and more closely resemble a bulk silicon transistor. Using the same maximum transconductance method, a maximum μ_{chan} of 34 and 31 cm^2/(Vs) was extracted for these WG and CG devices, respectively.

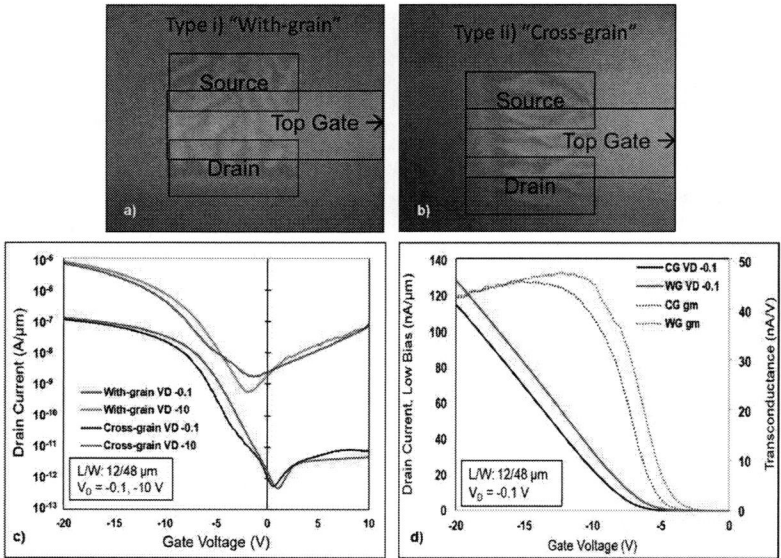

Figure 5. a) Micrograph of WG style structure, highlighting the parallel direction of grain boundaries in the channel. b) Micrograph of CG style structure, highlighting the perpendicular direction of grain boundaries in the channel. c) Transfer characteristic overlay of a typical pair of WG (red) and CG (black) transistors of the same dimensions and processing history. d) Linear scale low-drain behavior and transconductance calculation of the devices in section c.

Conclusions

Very thin layers of chromium can act as an effective adhesion layer during flash lamp annealing of amorphous silicon to enhance the wettability of the momentarily liquid silicon to glass substrates. When the thickness of this metal underlayer and the intensity of FLA pulse is properly calibrated, the chromium can act as both an adhesion promoter and a barrier between liquid silicon and borosilicate glass. In the course of this crystallization, an edge-directed polycrystalline morphology emerges rather than one filled with randomized voids. This grain morphology is predictable and can be used to produce TFTs with a high degree of uniformity, mitigating one of the main obstacles to an industrial integration of FLA.

This edge-oriented grain morphology allows for direct comparison between devices that are built with channels in the same direction as the grains and those that cut across it. It has been shown that CG devices have shallower switching characteristics and slightly reduced output current than WG devices. Unfortunately, issues with dopant introduction and dangling bond passivation complicate the analysis of this experiment; however, those can be easily rectified in the future. More analysis is being performed on the Cr-induced edge-directed grain structure itself to quantify grain size and orientation, which can be related to delivered FLA energy.

Further study will work towards understanding the mechanism and final structure of chromium in this procedure. If the underlayer survived as a metal in the final devices, it would likely form a conductive pathway below the channel material and render transistor behavior impossible. It is instead possible that chromium has formed an oxide or a silicide, both of which form Schottky contacts with p-type silicon that may prevent conduction. Analysis of bond energy will illuminate the final state of this critical layer and determine if there is significant Cr penetration into the channel.

Acknowledgments

The authors would like to acknowledge the support of the technical staff at the RIT Semiconductor & Microsystems Fabrication Laboratory (SMFL), the AMPrint facility at RIT, and the technical staff at Corning Incorporated. Financial support has been provided by Corning Incorporated and NYSTAR, through the New York State Center for Advanced Technology.

References

1. K. Ohdaira, N. Tomura, S. Ishii, K. Sawada, and H. Matsumura, *J. Non Cryst. Solids*, **358**, p. 2154-2158 (2012)
2. M. Park, Z. Vangelatos, Y. Rho, H. K. Park, J. Jang, and C. P. Grigoropoulos, *Thin Solid Films*, **696** (2020)
3. H. Kim, D. Y. Jeoung, S. Lee, and J. Jang, *IEEE Electron Device Lett.*, **40**, p. 411-414, (2019)
4. J. H. Hwang, H. J. Kim, B. K. Kim, W. B. Jin, Y. Kim, H. Chung, and S. Park, *Int. J. Therm. Sci.*, **91**, p. 1-11, (2015)
5. G. Packard, R. G. Manley, and K. D. Hirschman, *ECS Trans.*, **90**, p. 79-88 (2019)
6. T. Mudgal, K. Bhadrachalam, P. Bischoff, D. Cormier, R. G. Manley, and K. D. Hirschman, *ECS J. Solid State Sci. Technol.*, **6**, p. Q179-181 (2017)
7. G. Packard, A. Rosenfeld, K. Bhadrachalam, V. Garg, R. G. Manley, and K. D. Hirschman, *Thin Film Transistor Technologies 14 (TFTT 14)*, **86**, p. 57-72 (2018)
8. C. C. Tsai, Y. J. Lee, J. L. Wang, K. F. Wei, I. C. Lee, C. C. Chen, and H. C. Cheng, *Solid State Electron.*, **52**, p. 365-371 (2008)
9. S. Jin, Y. Choe, S. Lee, T. W. Kim, M. Mativenga, and J. Jang, *IEEE Electron Device Lett.*, **37**, p. 291-294 (2016)

140

Flash Lamp Annealed LTPS TFTs with ITO Bottom-Gate Structures

G. Packard [a], A. Rosenfeld [a], R.G. Manley [b], and K.D. Hirschman [a]

[a] Department of Electrical and Microelectronic Engineering, Rochester Institute of
Technology, Rochester, New York, 14623, USA
[b] Corning Incorporated, Science & Technology Research,
Corning, New York, 14831, USA

A study on bottom-gate PMOS TFTs built on flash lamp annealed
polycrystalline silicon is presented. As an alternative to top-gate
devices, bottom-gate TFTs offer potential benefits in dielectric-
semiconductor interface quality due to process integration details
of crystallization and defect passivation. Indium Tin Oxide was
used as the gate electrode material due to its attractive optical and
electrical properties, thermal stability, and compatibility with the
FLA process. Details of an experimental design used to investigate
combinations of FLA and furnace annealing for a-Si crystallization
and dopant activation processes will be discussed. ITO Bottom-
gate TFTs fabricated with FLA crystallization followed by boron
ion implantation and furnace activation exhibited superior
electrical characteristics in comparison to other treatments. A
comparison of electrical characteristics measured on bottom-gate
and double-gate devices is used to develop an interpretation of
defect effects and provide a qualitative assessment of interface
states.

Introduction

Low-Temperature Polycrystalline Silicon (LTPS) thin-film transistors (TFTs) are usually
built as a top-gate (TG) structure. This is due to the serious challenge presented by locally
crystallizing a thin layer of amorphous silicon that has regions of overlap with an
absorptive and thermally conductive gate electrode and areas of thermally insulating
glass. Early attempts to create LTPS devices quickly switched from the industry standard
bottom gate to the more attainable top gate, in which a gate structure is deposited
subsequent to crystallization (1). Despite this difficulty, bottom-gate (BG) devices offer
potential benefits in the dielectric-semiconductor interface quality, as well as being a
requirement for double-gate (DG) devices and enabling higher fill-factor pixel designs in
TFT display backplanes. BG TFTs have been realized with crystallization via excimer
laser annealing (ELA) with a demonstrated improvement over TG devices (2). This is of
particular interest for the development of LTPS devices that employ an alternative
crystallization technique, such as metal-induced lateral crystallization (3, 4) and
continuous-wave laser annealing (5, 6), motivated by potential advantages in large-panel
manufacturing.

FLA (Flash lamp annealing) has been explored as an alternative to ELA LTPS (7),
and has the potential to streamline and decrease the cost of fabrication while providing

applicable electrical performance. In this method, amorphous silicon is exposed to a single pulse of light emitted from a xenon flash lamp, causing melting and recrystallization in an area of cm^2 rather than nm^2 in less than a millisecond. FLA has shown promise in crystallizing silicon and activating dopants in both PFET and NFET TFTs (8). The low wettability of liquid silicon on most low-absorption materials has often resulted in a crystallized material made of randomized voids and pathways, as shown in Figure 1a, with the density of voids proportional to the amount of energy absorbed during the anneal. Nevertheless, FLA has been used to realize functional devices such as shown in Figure 1b (7). Investigations on alternative material morphology are in progress (9).

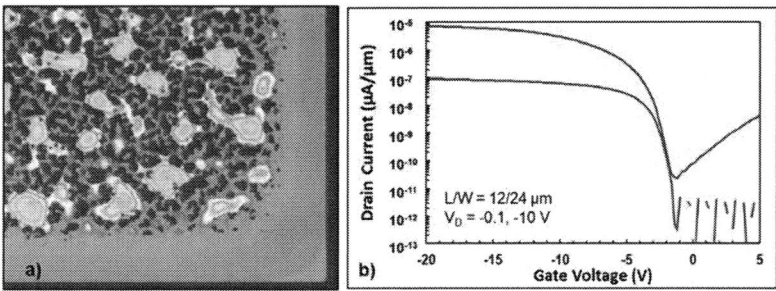

Figure 1. a) Micrograph of random silicon film dewetting in a mesa region following FLA crystallization, indicating characteristic voids and contiguous pathways, with optical artifacts of a capping oxide layer. b) Transfer characteristics of a PMOS TFT built on such material, $V_T = -2.8$ V, $\mu_{ch} = 44$ $cm^2/(Vs)$. (7)

Early attempts to produce BG FLA-LTPS TFTs were unsuccessful; the high degree of light absorption and reflectivity in metal or polysilicon gates resulted in severe, localized ablation of the overlying channel silicon. Reducing FLA emission intensity prevented this catastrophic channel loss, but also suppressed the crystallization of adjacent source/drain regions to the point where device performance was dominated by series resistance. Furthermore, the extent of this effect was heavily contingent on the area of the underlying gate features, as demonstrated in Figure 2. This geometric variance effectively precludes FLA-absorbent materials from functioning as bottom gates in this process method.

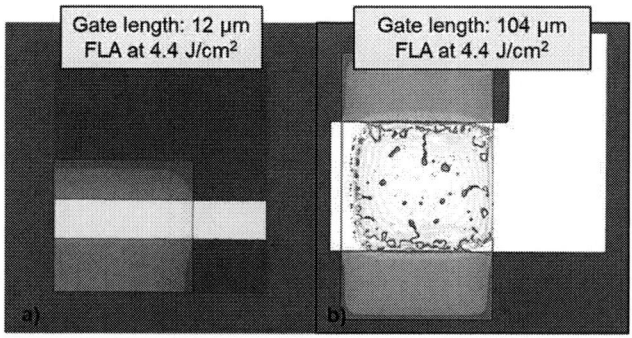

Figure 2. Demonstration of underlying molybdenum gate area influence on FLA using the same exposure conditions: 4.4 J/cm^2, 250 µs pulse, comparing gate lengths of L = 12 µm (a) and L = 104 µm. The smaller gate feature mesa does not achieve melt phase, whereas the larger gate feature experiences catastrophic silicon ablation.

Indium Tin Oxide (ITO) is commonly used in many electronics applications when a conductive material is required that is transparent in the visible light spectrum. These properties also make ITO potentially valuable as an alternative BG electrode material for FLA TFTs. By reducing the absorptivity of underlying materials, it may be possible to expose mesas of amorphous silicon to the flash energies necessary for broad crystallization without inadvertently causing catastrophic channel ablation. The incorporation of this material also supports an overall higher transmissivity.

Methods

This experiment used 150 mm Corning Lotus display glass substrates. A 40 nm ITO layer was deposited via sputter deposition directly from a ceramic target with no oxygen flow. This layer was patterned into bottom gate structures with HCl and annealed at 400 °C for two hours in air ambient to improve conductivity and resilience to chemistry. A 100 nm layer of PECVD SiO$_2$ was then deposited as a gate dielectric, followed by an active layer of 60 nm PECVD amorphous silicon, which was patterned into mesas with SF$_6$ plasma RIE. Another 100 nm PECVD SiO$_2$ was deposited as a capping and antireflective layer. The samples were then split into three batches with respect to doping, crystallization, and activation anneal, as shown in Table I.

TABLE I. Crystallization and Dopant Activation Experimental Design

Batch	Step 1	Step 2	Step 3
A	FLA crystallization	Boron implant	Furnace activation
B	Boron implant	FLA crystallization	--
C	Boron implant	FLA crystallization	Furnace activation

Batch-A was first crystallized using FLA, then received a boron ion implant, which was activated with a furnace anneal. Batches B and C received the boron implant first, followed by the FLA process to simultaneously crystallize a-Si and activate the implanted dopants. Batch-C received an additional furnace activation treatment, and substrates were apportioned to BG and DG devices. All FLA treatments used exposure energy densities of 4.2 − 4.5 J/cm^2 within 250 µs; all doping steps had 4×10^{15} cm^{-2} of ^{11}B$^+$ implanted at 35 keV through 100 nm screen oxide; all furnace anneals were done at 630 °C in N$_2$ for 12 hours. Samples were top-metallized with aluminum to produce both ITO BG and ITO/Al DG devices. Finally, devices were sintered at 450 °C in forming gas to passivate dangling bonds and promote ohmic contact formation.

Results

ITO Characteristics

The electrical and optical properties of indium tin oxide are heavily dependent on the conditions used for deposition and subsequent annealing. As deposited, sputtered 40 nm ITO films exhibited a sheet resistance of ~2100 Ω/sq with a transmission of between 0.49 and 0.75 in the visible range. However, an anneal of 400 °C for two hours in standard atmosphere reduced the sheet resistance to ~125 Ω/sq and increased transmission in the visible range to between 0.71 and 0.87. This amounts to an increase in the standard ITO figure of merit (FOM) by more than 19x, as shown in Figure 3. In addition, this anneal greatly increased the etch resistance of ITO to HCl, HF, and even SF_6 RIE.

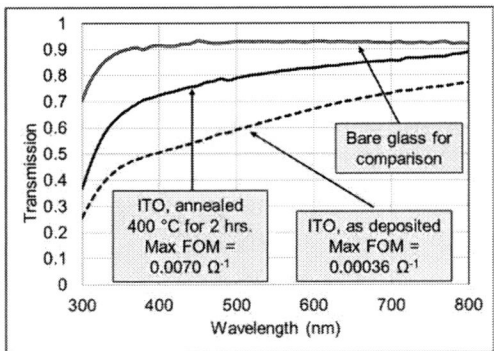

Figure 3. Transmission data of 40 nm ITO film as deposited (dashed line) vs after anneal (solid line), with bare glass substrate for comparison. Figure of merit: Transmission divided by average resistance, calculated at the 740 nm position.

Device Characteristics

BG transistors were fabricated using the three schemes depicted in Table I. Batch-A, in which silicon was crystallized prior to the introduction of dopants, offers the most control over the crystallization step. An example transistor is shown in Figure 4a and 4b, in the post-implant and completed stages, respectively. Distinct horizontal stripes are visible on the silicon mesa in Figure 4a, indicating regions that responded differently to process steps. The central bar is the eventual device channel, which was crystallized in the presence of an underlying ITO layer and was masked during the high-dose ion implant, thus retaining a high crystalline fraction. Narrow strips on both sides of the channel indicate regions crystallized above an ITO layer but not masked during implant, causing their crystal structure to become damaged by impinging ions. Beyond these strips lie outermost regions that were neither crystallized in the presence of ITO nor masked during implant; these regions are visually identical to unprocessed amorphous silicon.

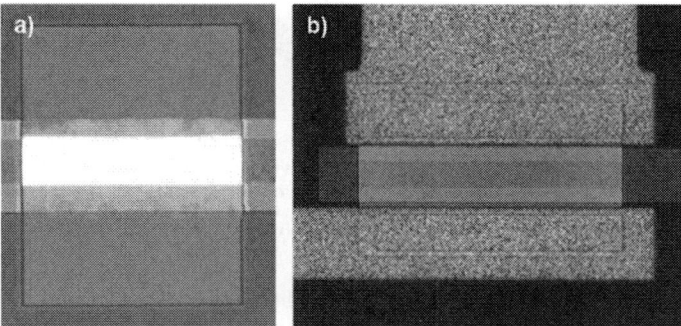

Figure 4. Micrographs of Batch-A TFTs with ITO bottom gates taken directly after ion implant (a) and as a completed device (b).

Despite the increased transparency of ITO, a-Si crystallized over an ITO underlayer during FLA still experiences an enhanced effective energy delivery when compared to outside regions. However, ITO is superior to opaque metals as a bottom gate in that this enhancement is not extreme enough to result in film ablation. Representative PFET characteristics from devices produced using this Batch-A process are shown in Figure 5.

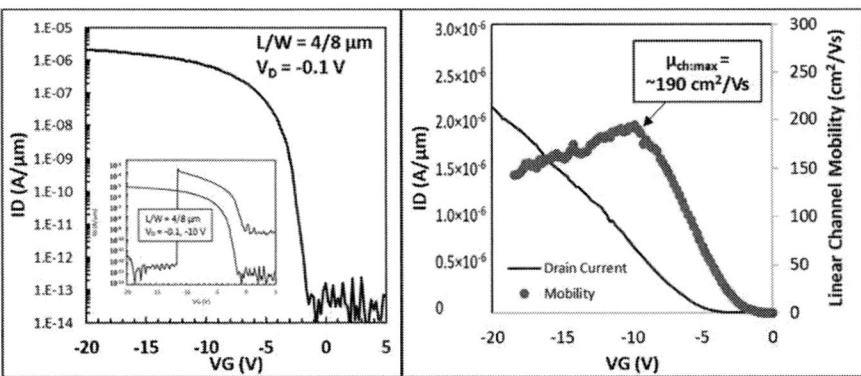

Figure 5. Transfer characteristic of a Batch-A transistor (FLA exposure $E = 4.4$ J/cm^2) with $L = 4$ μm and $W = 8$ μm, operating in low drain bias. The extrapolated threshold voltage $V_T \sim$ -4.8 V, and the extracted hole channel mobility $\mu_{ch} \sim$190 cm^2/(Vs) using the maximum transconductance method. The inset on the left shows destructive failure at $|V_G| < 12$ V at high drain bias.

These devices exhibited excellent electrical characteristics at low drain bias, with threshold voltages between 0 and -5 V and I_{ON}/I_{OFF} current ratios greater than 7 orders of magnitude (the parametric analyzer used here could not reliably measure $I < 100$ fA). Using the maximum transconductance method, a maximum linear-mode channel mobility was extracted as roughly 190 cm^2/(Vs), which is nearly three times higher than mobilities obtained from a standard TG device. This high value is likely a factor of the improved interface between the silicon and the gate dielectric, as it is established during the same localized melting process that forms the polycrystalline silicon itself, rather than being subsequently deposited on top. As a result, dangling silicon bonds at the gate dielectric interface are greatly reduced.

However interface defects still remain an issue, presumably at the back-channel region, as shown in the Figure 5 inset by the shifted characteristic at high drain bias. In addition, most of the Batch-A devices proved to be unexpectedly fragile when tested at high drain bias. While the off-state leakage remained relatively low (i.e. $I_{OFF} \sim 10^{-9}$ A/um at $V_D = -10$ V), the on-state reached a destructive breakdown at a gate voltage typically around the same value (i.e. $V_G \sim V_D = -10$ V), also shown in the Figure 5 inset. The failure mechanism is expected to be the severing of conductive channel pathways under high current / high field test conditions, which may be aggravated by the aggressive sinter process that allows for more silicon consumption by the contact aluminum. Another possible cause is gate dielectric failure in the G-S/D overlap regions due to residual damage caused by a significant fraction of the high fluence $^{11}B^+$ implant that ends up making it through the silicon mesa.

Batch-B devices received a single FLA exposure for dopant activation without furnace annealing. Figure 6a shows the transfer characteristics of a typical BG device, with the drain bias limited to -5 V to avoid destructive failure. Note that in comparison to the Batch-A device, the leakage current is over two orders of magnitude higher at half the drain bias, with a very low I_{ON}/I_{OFF} ratio. Very few devices with $L \leq 8$ μm displayed any switching behavior. Although the structure of these devices limits the molten region of each mesa to that directly above the ITO features, results suggests that boron at the channel edge may segregate into the liquid phase and migrate some distance into the channel, rendering the transistor channel length significantly shorter than designed.

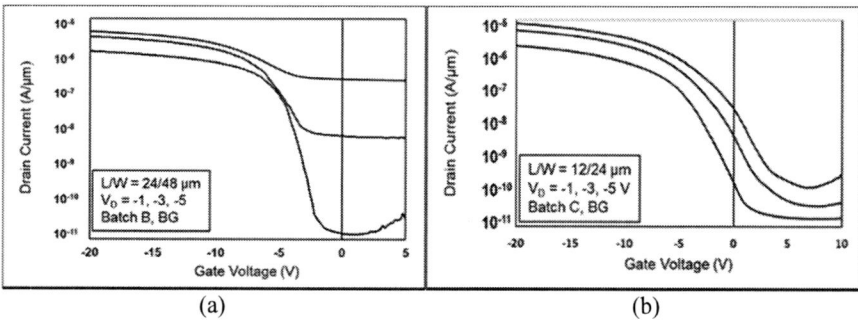

(a)   (b)

Figure 6. a) Transfer characteristics of a Batch-B transistor (L/W = 24/48 μm) processed with a single FLA exposure at 4.5 J/cm², without a furnace anneal. b) Transfer characteristics of a Batch-C transistor (L/W = 12/24 μm) processed with the same FLA exposure, with the addition of a furnace activation anneal. Both devices were tested with drain bias limited to $|V_D| \leq 5$ V.

Figure 6b shows a representative device from Batch-C, in which the FLA crystallization/activation step was followed by a standard furnace activation at 630 °C for 12 hours. This secondary process was incorporated into the experiment in order to ensure adequate dopant activation. The additional furnace anneal appears to have helped mitigate the pronounced off-state leakage shown by the Batch-B device, at the expense of a reduced subthreshold slope and increased characteristic separation shown by the Batch-C device. This may be due to the involvement of competing mechanisms; the reduction of a primary defect effect that remains after FLA may be responsible for the promotion of a secondary defect effect that is enhanced during annealing.

Some devices in Batch-C were finished as DG transistors for a comparison of electrostatic effects. These transistors coupled an ITO bottom gate with an aluminum top gate for fabrication simplicity. Such a pairing is possible because the top gate need not be compatible with the FLA exposure process. Figure 8 demonstrates an array of switching characteristics obtained from a single Batch-C device when operated in different ways; with the gate electrodes linked as a true DG device, or with one of the gate electrodes held at reference ground while sweeping the other.

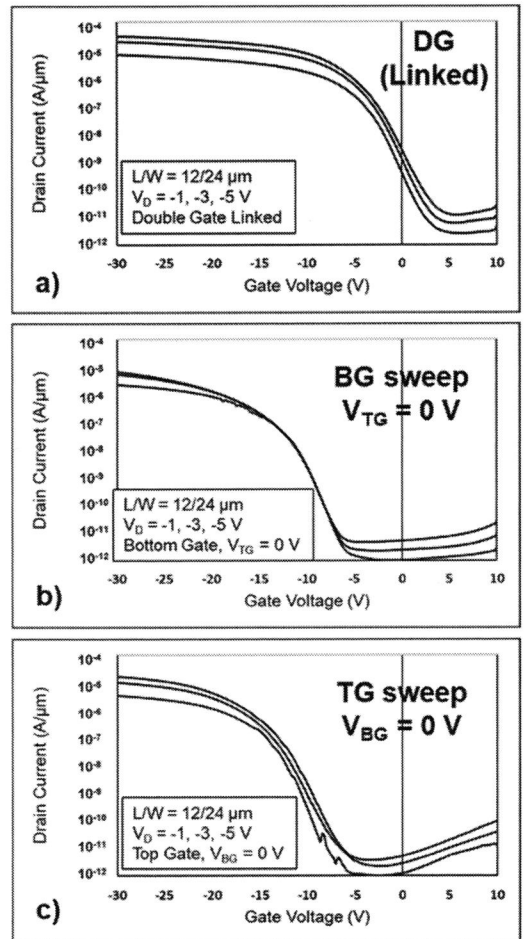

Figure 7. A single Batch-C TFT tested in three different control modes: a) DG (linked), b) BG ($V_{TG} = 0$ V), c) TG ($V_{BG} = 0$ V). In all cases, the drain bias was limited to $|V_D| \leq 5$ V.

Similarities and differences in transfer characteristic behavior shown in Figure 7 provide a certain degree of insight on the nature of the FLA LTPS channel and associated oxide interfaces. The linked DG configuration represented in Figure 7a provides more current drive and a steeper subthreshold slope in comparison to the BG device in Figure 6b, due to improved electrostatic control on channel charge. The transfer characteristic separation remains, though reduced significantly. The single-gate control mode characteristics are shown in Figures 7b and 7c. The grounded electrode is constantly fighting against the control gate modulation of free channel charge, and influences the effective level of trapped charge in defect states as well. This combination results in the significant left-shift in the transfer characteristic, as well as degradation in the subthreshold slope. The BG control mode yields less current than the TG control mode, which is likely due to asymmetries such as the respective oxide thickness and gate electrode workfunction. A more obvious difference is the separation in the transfer characteristics, which appears to vanish in the BG-control mode of operation. This suggests an asymmetry in trap states at the respective gate dielectric interface. With V_{TG} = 0 V, interface traps at the top dielectric interface remain predominantly fixed while interface traps at the bottom dielectric interface respond to the BG voltage. With $V_{BG} = 0$ V the situation is vice versa; interface traps at the bottom dielectric interface remain fixed while interface traps at the top dielectric interface respond to the TG voltage. The overlay in the BG-control mode characteristics infers that the bottom dielectric interface is superior to the top dielectric interface. This is consistent with the presumption previously stated in describing the behavior of the Batch-A BG device.

Summary & Conclusions

The results of the experimental design used to investigate treatment combinations of FLA and furnace annealing for a-Si crystallization and dopant activation processes provided important understanding on the mechanisms involved with establishing BG TFT operating behavior. Batch-B devices demonstrated elevated off-state leakage, which appears to be associated with defects that form during and remain after the FLA process with boron dopant atoms present. The leakage current seems to improve if FLA activation is followed by a furnace anneal as shown by Batch-C device behavior; however a secondary defect effect seems to arise which compromises the subthreshold slope. Batch-A treatments which had FLA for mesa crystallization and furnace annealing for dopant activation showed the most promise, taking advantage of the superior BG dielectric interface and avoiding some of the defect issues present in Batch-B and Batch-C treatments. However, Batch-A TFTs exhibited behavior associated with back-channel defects at the passivation oxide interface, as well as premature failure under high current / high field test conditions. Note that this failure mode was avoided when testing Batch-B and Batch-C devices by enforcing a conservative drain bias limit.

The multi-mode testing of a Batch-C DG TFT provided additional insight on the back-channel issue experienced by Batch-A BG devices. The DG test results confirmed that, as expected, the bottom dielectric interface was superior to the top dielectric interface. The top dielectric interface of the DG device is exactly the back-channel interface of the BG device. The quality of the back-channel interface may be improved by leaving the capping oxide that is used for the FLA process intact, rather than removing it following the $^{11}B^+$ implant and FLA processes and depositing a replacement PECVD oxide layer. The original capping oxide interface should be similar in quality as the bottom dielectric interface, as they are formed simultaneously during the FLA process. Further improvement may be realized using alternative defect passivation strategies.

Acknowledgments

The authors would like to acknowledge the support of the technical staff at the RIT Semiconductor & Microsystems Fabrication Laboratory (SMFL), the AMPrint facility at RIT, and the technical staff at Corning Incorporated. Financial support has been provided by Corning Incorporated and NYSTAR, through the New York State Center for Advanced Technology.

References

1. Y. Miyata, M. Furuta, T. Yoshioka, and T. Kawamura, *Jpn. J. Appl. Phys.*, **31**, (12S), 4559, (1992)
2. K. Oh, S. Yang, J. Lee, K. Park, and M.Y. Sung, *Electron. Lett.*, **51** (24), 2030 (2015)
3. J.H. Oh, K.W. Ahn, D.H. Kang, W.H. Park and J. Jang, *Solid State Electron*, **52** (3), 482 (2008)
4. S.K. Lee, K.H. Seok, J.H. Park, H.Y. Kim, H.J. Chae, G.S. Jang, Y.H. Lee, J.S. Han, and S.K. Joo, *Appl. Phys. A*, **122**, 613 (2016)
5. A. Hara, M. Takei, K. Yoshino, F. Takeuchi, M. Chida, and N. Sasaki, *IEEE International Electron Devices Meeting (IEDM)*, 8.6.1, (2003)
6. A. Hara, T. Sato, K. Kondo, K. Hirose, and K. Kitahara, *Jpn. J. Appl. Phys.*, **50** (2R), 021401 (2011)
7. G. Packard, A. Rosenfeld, K. Bhadrachalam, V. Garg, R. G. Manley, and K. D. Hirschman, *ECS Trans.*, **86** (11), 57 (2018)
8. T. Mudgal, K. Bhadrachalam, P. Bischoff, D. Cormier, R. G. Manley, and K. D. Hirschman, *ECS J. Solid State Sci. Technol.*, **6**, Q179 (2017)
9. G. Packard, A. Rosenfeld, M. Hum, R.G. Manley, and K.D. Hirschman, *ECS Trans., TFT 15, this issue* (2020)

Photoconductive Solution Processed ZnO Quasi-superlattice Films

Darragh Buckley[1], Saikumar Inguva[1], David McNulty[1], Vitaly Zubialevich[2], Peter Parbrook[2,3], Farzan Gity[2], Paul Hurley[1,2,4], and Colm O'Dwyer[1,2,4,5]

[1]School of Chemistry, University College Cork, Cork, T12 YN60, Ireland
[2]Tyndall National Institute, Lee Maltings, Cork, T12 R5CP, Ireland
[3]School of Engineering, University College Cork, Cork, T12 YN60, Ireland
[4]AMBER@CRANN, Trinity College Dublin, Dublin 2, Ireland
[5]Environmental Research Institute, University College Cork, Lee Road, Cork T23 XE10, Ireland

> Photoconductance is reported from single layer and 20 layer quasi-superlattice films of ZnO grown by solution processing and spin coating. At sub-band (532 nm) excitation, significant and stable photoconducrrent is observed and can be modulted by laser power density.

INTRODUCTION

Metal oxide TFTs have a greater performance and stability in TFTs than amorphous silicon devices and are more applicable to transparent, flexible displays which are the foundation of next generation, interactive displays.(1, 2) From the metal oxides, zinc oxide (ZnO)-based materials have attracted increasing attention for use in flexible displays because of their higher mobility and lower processing temperature than conventional hydrogenated amorphous Si (a-Si:H) TFTs.(3, 4) Zinc oxide is an material of interest in the area of optoelectronics due to its wide band gap ($E_g \sim 3.3$ eV at 300 K)(5), large exciton binding energy (~66 meV) and especially for the variety of methods by which it can be processed(6). Moreover, ZnO is inexpensive, abundant and readily able to alloy with other metals in the oxide form(7-9) and has a lattice that can facilitate interstitial doping.(4) This gives ZnO a key role in the area of optoelectronics, metal oxide thin films and thin film transistor (TFT) technologies.(10, 11)

Solution processing has shown to be very useful for developing homogeneous quasi-superlattice films, and even hetero-interfaces between oxide that permit high mobility FET devices and thin film transistors(2, 12-16). The knowledge of electronic properties and structure of superlattice oxides is important to understand the nature of electronic structure and how it affects and controls optoelectronic characteristics of FETs, TFT, photodetectors and other optoelectronic devices.

Here, we report that photoconductance is possible from ZnO films and quasi-superlattices grown by solution processing when excited at sub-band gap energies. At 532 nm, we show repeatable and consistent photoconductivity at 532 nm, but no band-to-band recombination leading to photoluminescence is observed at the same excitation energy. Under bias, ZnO quasi-superlattice films with high crystallinity becomes relative good photoconductors.

EXPERIMENTAL

ZnO quasi-superlattice film growth

Zinc oxide thin films were synthesised from zinc acetate dihydrate [$Zn(CH_3COO)_2.2H_2O$]. The acetate powder was dissolved in 2-methoxyethanol [$CH_3OCH_2CH_2OH$] to make a 0.75 M solution. An equimolar solution of monoethanolamine (MEA) was added to the zinc precursor as a stabilising agent. All chemicals were purchased from Sigma Aldrich. The precursor was heated at 60 °C for 2 h and stirred until a clear, colourless solution is obtained. Controllably thin and smooth films were prepared by a spin-coating technique. Using a SCS G3 desktop spin coater, ZnO precursor solution was added dropwise to cover substrates and were spun at 3000 rpm for 30 s, including a 5 s ramp time. Thin films were deposited on doped p-type silicon substrates that were covered with a 300 nm thick layer of thermally grown SiO_2. Substrates for spin coated samples were cut into square coupons of nominal dimension 2 cm × 2 cm to ensure even coverage as the samples are rotated about a perpendicular symmetric axis(17, 18). An acetone, IPA and DI water wash combined with sonication was used to ensure that the surface of the wafers were clean of surface contaminants prior to thin film deposition. A UV-Ozone treatment was performed for 30 mins using a Novascan UV ozone system to further remove any organic contaminants from the surface. After a layer of precursor was spin coated, samples were dried in an open-air convection oven at temperatures between 250 - 270 °C for 5 min. Another layer of liquid precursor was then spun onto the surface followed by more oven-drying. This process was then repeated a number of times to acquire the desired number of deposited layers and the samples were subsequently annealed at 300 °C for 1 hr as a final heat treatment.

Physical and structural characterization

Transmission electron microscopy (TEM) was conducted on lamellar cross-section of the ZnO thin films on SiO_2/Si. Cross-sectioning of the Si electrodes was carried out with an FEI Helios Nanolab Dual Beam FIB System. Cross-sectional TEM sample preparation was performed on slices using a standard FIB lift-out technique. TEM analysis and electron diffraction was conducted using a JEOL JEM-2100 TEM operating at 200 kV. X-ray diffraction (XRD) was used to characterise the crystallographic structure of the single and multi-layer ZnO films after spin coating deposition. XRD analysis of ZnO films was performed using a Philips X'Pert PW3719 diffractometer using Cu Kα radiation (40 kV and 35 mA) scanned between 10 - 80° (2θ).

Electrical and spectroscopic measurements

Electrical transport characterisation was carried out using a 2-probe method on samples in the dark and under laser illumination. I-V data were recorded using a Keithley 2600B sourcemeter with 50 ms per point integration time. For photoconductance measurements, films were contacted with indium-gallium eutectic to form contacts with the AZO thin films. Samples were tested under dark conditions and under laser excitation. Illumination at 532 nm was provided by a Laser Quantum GEMS diode pumped solid state laser with power varying from 5 to 80 mW, generating power densities at the sample surface ranging from 1.6 to 25.5 mW cm^{-2}. Hall-effect measurements were performed in a van der Pauw configuration with a LakeShore Model 8404 AC/DC Hall Effect Measurement System. The system can provide DC or AC magnetic fields over a variable range up to

±1.7 T (DC) or a fixed range up to ~1.2 T RMS at a frequency of 0.1 Hz or 0.05 Hz. The measurements in this work were performed at room temperature.

Photoluminescence (PL) spectroscopy was used to probe the electronic structure of the spin coated ZnO thin films. Photoluminescence (PL) spectroscopy was used to probe the electronic structure of the spin-coated thin films. Band-edge PL emission was carried out at room temperature using a 325 nm He–Cd laser excitation source with power density of 2 W/cm^2. PL spectra were recorded using a Horiba iHR320 spectrometer equipped with a thermoelectrically cooled Synapse CCD matrix. Low-temperature photoluminescence (PL) spectra were recorded using He–Cd laser excitation at 325 nm with a 1 m SPEX 1704 monochromator that coupled to a Hamamatsu model R3310-02 photo-multiplier tube. Excitation at 532 nm was provided by a Laser Quantum GEMS diode pumped solid state laser also coupled to and Horiba iHR320 as above. The monochromator contained a grating blazed at 330 nm (ISA model 510-05). All samples were loaded in a cryostat (Janis model SHI-950-5), which was cooled to 11 K using a closed cycle helium-gas refrigerator.

RESULTS AND DISCUSSION

Focused ion beam-thinned lamella of deposited ZnO thin films were examined via TEM to determine the internal structure and to investigate if an iterative spin-coating deposition scheme results in a continuous film or periodic lattice formation. The TEM analysis shown in Fig. 1 shows the cross-sectional structure of a single-layer ZnO thin film and a multi-layer quasi-superlattice (QSL) consisting of 20 individually spin-coated layers. The first deposited layer is shown to uniform in thickness across the SiO_2 substrate. Figure 1(b) displays the layered structure of the multi-layered QSL. Subsequent deposition of material does not interfere with the uniformity of this initial layer.

In this periodic structure, each deposition results in a bilayer comprising a dense ZnO material atop a granular sublayer. The drying step (5 min at 260 °C) between spin-coats results in the formation of the denser capping layer on the upper surface of each deposition, while the sub-layer beneath appears more granular and porous in TEM analysis. This bilayer structure is credited with the epitaxial-like growth of these thin films on amorphous substrates(19). The resulting QSL or single-layer thin film is crystallised during the final annealing treatment (1 hr at 300 °C). There is a distinct change in the texture of the crystal growth habit during the sequential process of spin coating additional precursor material and the associated heat treatments. The 20 layer QSL displays a growth preference to the [0002] c-plane direction which is due to the dense capping layer induced by the drying treatments after each spin-coat. The partially porous layer between these iterative capping layers corresponds to the m-plane reflections still present in the multi-layered film's diffraction pattern. The absence of the other reflections seen in the polycrystalline ZnO pattern acquired from a heated drop-cast powder (inset, Fig. 1(c)) confirms an epitaxial-like growth of spin coated films on an amorphous substrate. This suggests that the initial, dried layer of ZnO on the SiO_2 forms a template for the directed growth of multi-layer or QSL films; subsequent growth however, and cumulative annealing converts the crystal texture to predominantly c-plane orientation.

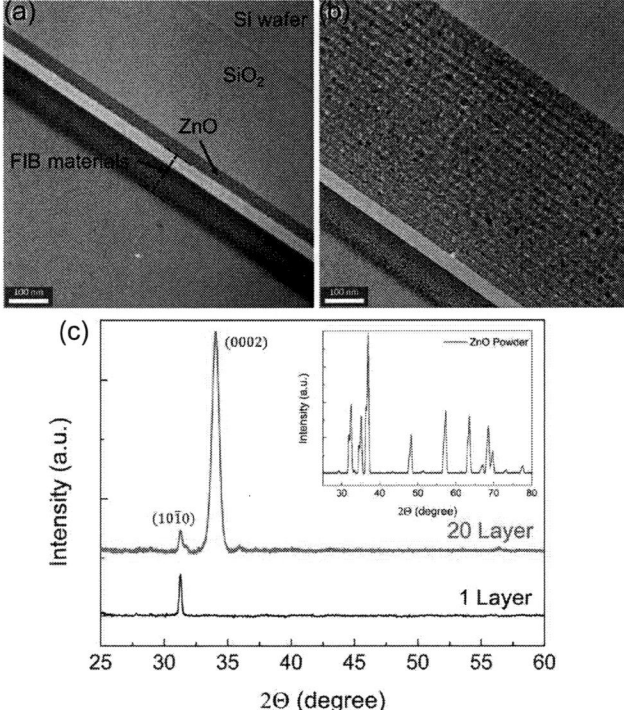

Figure 1. TEM cross-sectional images of (a) 1 layer and (b) 20 layer ZnO thin films grown on SiO₂/Si substrates. The FIB processing materials comprise carbon and Pt straps. (c) X-ray diffraction patterns for 1 and 20 layer ZnO thin films. Inset is the diffraction pattern for pure ZnO powder.

ZnO is well known as a wide bandgap semiconductor, and to some extent is considered a reasonably good photodetector material, or at least its ability to absorb UV photons can occur in tandem with electrical transport measurements. ZnO has the smallest exciton binding energy of any semiconductor, and as such photogenerated carriers can be extracted to an outer circuit with high efficiency without deleterious thermal decomposition of bound excitons in ZnO. Even in the UV range at super bandgap energies, the high exciton binding energy renders ZnO a good photoconducting material. Even while ZnO is ubiquitous in many research fields, the standard approach to photoexcitation is to use super bandgap excitation to ensure valence band-to-conduction band electron pumping. As ZnO can be quite defective, and often produced with very small quantities of Al, Sn or other impurities, we investigated sub-band pumping to examine the defect levels of our QSL films. This uncovered a photoconductance phenomenon in ZnO illumination at 532 nm, i.e. sub band pumping produces a very stable, photoconductive ZnO film and quasi-superlattice layer. To enhance photoconductance, or indeed to enable sub-band gap photoconductivity in ZnO, excitonic binding remains crucial. Any photogenerated carrier will leave an electron hole resulting in an electric field that promotes recombination, which is why ZnO as a photoconductor material typically requires photon energies greater than the bandgap and is somewhat less sensitive to photon flux compared to other compound semiconductors, especially for room-temperature operation.

Figure 2 shows I-V testing carried out under various conditions of illumination to determine the optoelectronic properties of the spin-coated thin films. I-V data are shown in Fig. 3(b,c) from the 1 and 20 layer ZnO film form 0 – 40 V, respectively. I-V curves are shown when acquired under dark conditions and also under collimated illumination. Under illumination, we acquired the I-V curves by incrementing the laser power as follows: 10, 20, 40 and 80 mW and then in decrements from 80 to 10 mW, with a dark conditions curve acquired last. This sequence of tests are illustrated in the inset of Fig. 7(b). This allowed for the repeatability and consistency of the photoconductance to be examined.

These solution-deposited ZnO thin films clearly provide a similar photocurrent response in dark conditions immediately after being tested at higher laser powers demonstrates the remarkable optoelectronic stability of these films. The I-V characteristic typically appear ohmic, with a transition to non-linear photoconductance at higher laser powers. The I-V response in all cases is non-linear, and we observe a photocurrent saturation >20 V at the lowest illumination power of 10 mW for 20 layer ZnO QSL films, and any energy barrier to photocurrent enhancement occurs once the photon flux is increased. The ZnO QSL films are stable against focused high photon flux at 80 mW power and associated photothermal effects, and the ZnO films recover to a less-excited state without the requirement for time delay or passivation. We emphasize here that each pair of I-V curves at similar laser power, are acquired before and after all other tests to higher laser powers, and so these ZnO QSL are very stable during sub-band illumination.

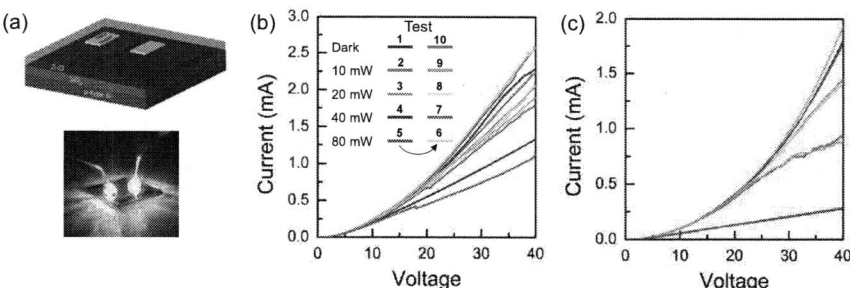

Figure 2. (a) Schematic and image of the 2-probe photoconductance measurements using In-Ga eutectic ohmic contacts to the top surface of the ZnO films with a 5 mm channel length between source and drain contacts. (b) I-V response of 1 layer and (c) 20 layer ZnO QSL films under 532 nm irradiation.

Some figures of merit for photoconducting materials were also examined and compared to other examples in the literature, including those where super bandgap illumination was used. Responsivity of a photoconductor, R_s, is calculated according to $R_s = \frac{I_{ph}}{E_W A_C}$, where $E_W A_C$ represent the optical power applied to the sample surface.

Figure 3 shows the measured photocurrent and responsivity of sub-band pumped ZnO single and 20 layer QSL films. We find 20 L films have a higher photocurrent, due to a smaller dark current compared to single layer films (*cf.* Fig 2).

The responsivity compared well to the literature, however known values were acquired under UV excitation. Here, clearly an electron donor defect level provide electrons to the conduction band, without leaving a hole in the valence band, allowing extraction of current under sub-band illumination. The responsivity here is based on full illumination of the 2 cm × 2cm sample, and can be increased by focusing the illumination on the channel region, which will be investigated in the future.

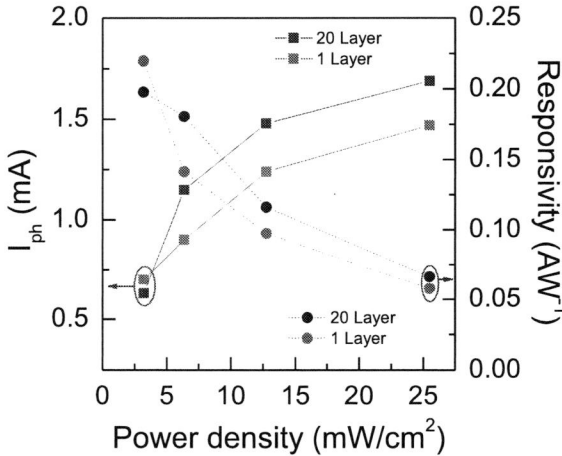

Figure 3. Maximum photocurrent measured at 40 V for 1 layer and 20 layer quasi-superlattice ZnO films and the corresponding responsivity determined over the entire area of the film on the 2 cm × 2 cm substrate.

To examine the nature of photoconductance in ZnO QSL films, we examined the structure using photoluminescence measurements with super bandgap (325 nm HeCd laser) and sub-bandgap (532 nm SS laser) incident photon energies. These measurements were acquired to explore and compare photoluminescence from ZnO at room temperature, and compare to photoconductance data. Bandgap photoluminescence is observed at ~375 nm (3.3 eV) for 1 and 20 layer ZnO when excited using 325nm laser excitation. These emissions are expected from stoichiometric ZnO. This band edge peak comprising band-to-band recombination and donor-accepter (DAP) pair emission, is present for both the 1 layer and 20 layer ZnO, with some minor shift observed in peak position.

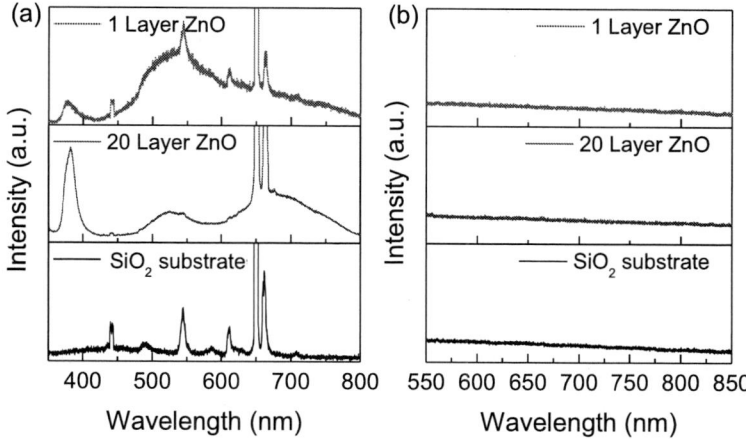

Figure 4. Photoluminescence spectra for 1 and 20 layer AZO and ZnO thin films, acquired using (a) $\lambda = 325$ nm and (b) $\lambda = 532$ nm excitation sources. PL spectra of SiO_2 substrates are included to highlight peaks attributed to background light sources and second order laser harmonics at $\lambda = 650$ nm and $\lambda = 1064$ nm, respectively.

Sub-band defect emission is also noted for the ZnO under 325 nm excitation, due to radiative recombination via electronic defects in the ZnO crystal lattice. For films probed using 532 nm radiation, no PL is noted, indicating that promotion of electrons from the valence band to the conduction band is necessary for PL of ZnO (band edge and intraband defect emission) when solution-deposited as a thin film or QSL. This latter observation is consistent with defect PL from ZnO arising from recombination with electron holes, with electrons located in shallow intraband defect levels. We consistently observe stable and strong photocurrents from these films when excited at 532 nm when driven by an electric field under source-drain bias.

CONCLUSIONS

We detailed the growth of crystalline ZnO thin films and quasi superlattices on amorphous substrates with a periodic layered structure from a solution-based precursor via a spin-coating technique. Electron microscopy and X-ray diffraction showed a periodic, epitaxial-like growth for multilayered films with a highly-ordered crystalline growth direction. While PL show typical response associated with valance band-conduction band emission, we show that reproductible and stable photoconductance is observed in single and 20 layer quasi-superlattice ZnO under sub-band 532 nm excitation that is modulated with laser power density.

ACKNOWLEDGEMENTS

We acknowledge support from the Irish Research Council Government of Ireland Postgraduate Scholarship under award no. GOIPG/2014/206. This work was supported by Science Foundation Ireland under contract no. 17/TIDA/4996. This publication has also emanated from research supported in part by a research grant from SFI under Grant Number 14/IA/2581.

REFERENCES

1. S. Shigehiko, H. Takeo, K. Motoki, K. Kazuto, Y. Mitsuaki and I. Masataka, *Jpn. J. Appl. Phys* **47**, 2845 (2008).
2. J. Socratous, K. K. Banger, Y. Vaynzof, A. Sadhanala, A. D. Brown, A. Sepe, U. Steiner and H. Sirringhaus, *Adv. Funct. Mater.*, **25**, 1873 (2015).
3. H. Lingling, H. Dedong, C. Zhuofa, C. Yingying, W. Jing, Z. Nannan, D. Junchen, Z. Feilong, L. Lifeng, Z. Shengdong, Z. Xing and W. Yi, *Jpn. J. Appl. Phys*, **54**, 04DJ07 (2015).
4. M. Lorenz, M. S. R. Rao, T. Venkatesan, E. Fortunato, P. Barquinha, R. Branquinho, D. Salgueiro, R. Martins, E. Carlos, A. Liu, F. K. Shan, M. Grundmann, H. Boschker, J. Mukherjee, M. Priyadarshini, N. DasGupta, D. J. Rogers, F. H. Teherani, E. V. Sandana, P. Bove, K. Rietwyk, A. Zaban, A. Veziridis, A. Weidenkaff, M. Muralidhar, M. Murakami, S. Abel, J. Fompeyrine, J. Zuniga-Perez, R. Ramesh, N. A. Spaldin, S. Ostanin, V. Borisov, I. Mertig, V. Lazenka, G. Srinivasan, W. Prellier, M. Uchida, M. Kawasaki, R. Pentcheva, P. Gegenwart, F. M. Granozio, J. Fontcuberta and N. Pryds, *J. Phys. D: Appl. Phys.*, **49**, 433001 (2016).
5. Ü. Özgür, Y. I. Alivov, C. Liu, A. Teke, M. A. Reshchikov, S. Doğan, V. Avrutin, S.-J. Cho and H. Morkoç, *J. Appl. Phys.*, **98**, 041301 (2005).
6. E. Fortunato, P. Barquinha, A. Pimentel, A. Gonçalves, A. Marques, L. Pereira and R. Martins, *Thin Solid Films*, **487**, 205 (2005).
7. A. Lyubchyk, A. Vicente, P. U. Alves, B. Catela, B. Soule, T. Mateus, M. J. Mendes, H. Águas, E. Fortunato and R. Martins, *Phys. Status Solidi A*, **213**, 2317 (2016).
8. R. Martins, P. Barquinha, A. Pimentel, L. Pereira and E. Fortunato, *Phys. Status Solidi A*, **202**, R95 (2005).
9. P. Nunes, E. Fortunato, P. Tonello, F. Braz Fernandes, P. Vilarinho and R. Martins, *Vacuum*, **64**, 281 (2002).
10. E. M. C. Fortunato, P. M. C. Barquinha, A. C. M. B. G. Pimentel, A. M. F. Gonçalves, A. J. S. Marques, L. M. N. Pereira and R. F. P. Martins, *Adv. Mater.*, **17**, 590 (2005).
11. C. Glynn and C. O'Dwyer, *Adv. Mater. Interfaces*, **4**, 1600610 (2017).
12. H. Faber, S. Das, Y.-H. Lin, N. Pliatsikas, K. Zhao, T. Kehagias, G. Dimitrakopulos, A. Amassian, P. A. Patsalas and T. D. Anthopoulos, *Sci. Adv.*, **3** (2017).
13. Y.-H. Lin, S. R. Thomas, H. Faber, R. Li, M. A. McLachlan, P. A. Patsalas and T. D. Anthopoulos, *Adv. Electron. Mater.*, **2**, n/a (2016).
14. Y. H. Lin, H. Faber, J. G. Labram, E. Stratakis, L. Sygellou, E. Kymakis, N. A. Hastas, R. Li, K. Zhao, A. Amassian, N. D. Treat, M. McLachlan and T. D. Anthopoulos, *Adv. Sci.*, **2**, 1500058 (2015).
15. S. R. Thomas, P. Pattanasattayavong and T. D. Anthopoulos, *Chem. Soc. Rev.*, **42**, 6910 (2013).
16. T. Schneller, R. Waser, M. Kosec and D. Payne, *Chemical Solution Deposition of Functional Oxide Thin Films*, p. 796, Springer, London (2013).
17. D. Buckley, D. McNulty, V. Z. Zubialevich, P. J. Parbrook and C. O'Dwyer, *ECS Trans.*, **77**, 99 (2017).
18. D. Buckley, R. McCormack, D. McNulty, V. Z. Zubialevich, P. J. Parbrook and C. O'Dwyer, *ECS Trans.*, **77**, 75 (2017).
19. D. Buckley, R. McCormack and C. O'Dwyer, *J. Phys. D: Appl. Phys.*, **50**, 16TL01 (2017).

Chapter 4

H03 – Non-Si and Non-Oxide TFTs

Percolation Carbon Nanotube Thin Film Transistors

M. Shur, J. Park, Y. Zhang, X. Liu, and T. Ytterdal [a]

[a] Rensselaer Polytechnic Institute, Troy NY 12081, USA
[b] Electronics of the Future, Inc., Vienna, VA 22181, USA
[c] University of Trondheim, Norway

Carbon Nanotube (CNT) Thin Film Transistors use the CNT mats forming a random percolating network. The conduction contacts between the individual CNTs are determined by tunneling and are strongly affected by external agents, such as gases or biological fluids. The asymmetry of the contacts to each CNT enables the rectification of electromagnetic radiation impacted on the CNT mat with giant relative changes near the percolation point. Our experimental data show that sub-THz radiation shifts the percolation point at high intensity, when the rectified radiation changes the shape of the potential barrier between the CNT in a proximity contact. We now propose to operate the CNT Thin Film Transistors near the percolation point, where the gate bias shifts the percolation point by changing the shape of the CNT contact tunneling barriers. This ensures a sharp decrease of the subthreshold current slope.

Introduction

Excellent materials properties of carbon nanotubes (CNTs) could enable multiple potential applications including flexible interconnects [1], terahertz (THz) power limiters [2], high speed microwave [3] and digital electronic devices and circuits [4]. This paper reviews the CNT Field Effect Transistor (FET) compact models and focuses on the CNT mat properties near the percolation point that could be used to reduce the CNT Thin Film Transistor (TFT) subthreshold slope. The paper is organized as follows. First, we briefly review the emerging applications of the CNT technology. Then we introduce the compact model for a CNT FET and apply this model to investigate the CNT FET response in the terahertz (THz) range. In the next section, we discuss the properties of the CNT percolating mats. We recently reported on the nonlinear sub-THz transmission of the CNT mats linked to the THz radiation rectification due to the CNT contact asymmetry. We now propose to use this contact effect to reduce the subthreshold slope of the CNT TFT current-voltage characteristics.

CNT applications

The CNT applications include high performance transistors, beyond-silicon microprocessors, nanoscale plasmonic THz detectors, low-noise RF amplifiers, flexible and wearable devices, and flexible interconnects. Fig. 1 (from [5]) shows a CNT 16-bit microprocessor integrating over 14,000 CNTFETs on a 6.912 mm^2 die area. Fig. 2 (from [6]) shows an example of flexible high-performance carbon nanotube integrated circuits.

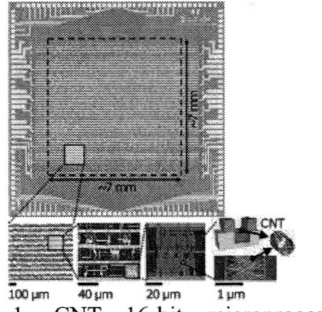

Fig. 1. CNT 16-bit microprocessor integrating over 14,000 CNTFETs. [5]

Fig. 2. Flexible high-performance carbon nanotube integrated circuit.[6]

Fig. 3 (from [7]) shows the application of a CNT for fluid delivery to a biological cell. Fig. 4 shows how CNT mats incorporated into polymer/metal composite improve the reliability of flexible interconnects. [1]

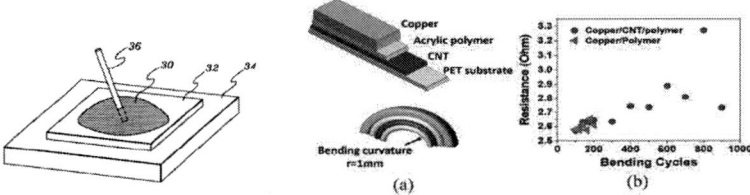

Fig. 3. CNT (36) used for fluid delivery to a biological cell. [7]

Fig. 4. CNT mats incorporated into polymer/metal composite for improving flexible interconnects.[1]

Compact model for a CNT field effect transistors and terahertz response

Fig. 5 shows the different CNT FET structures.

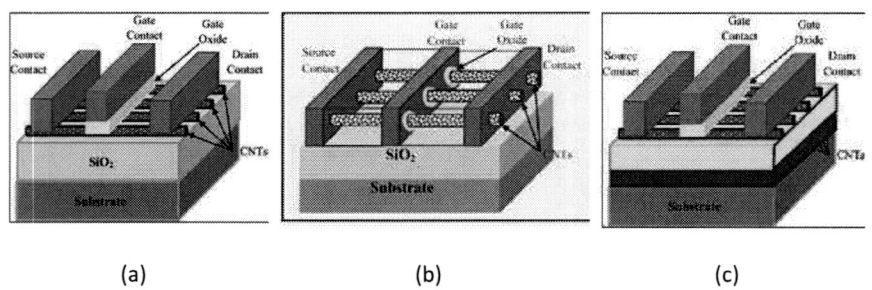

(a)   (b)   (c)

Fig. 5. Different CNT FET structures: (a) top gate CNT FET and (b) gate wrapped around CNT FET [8]; (c) top and bottom gate CNT FET.

The compact model of a CNT FET reported in [9] uses the Unified Charge Control model (UCCM) approach. In the frame of this model, the current-voltage characteristics of the CNTFET are given by

$$I_{ds} = \frac{g_{ch}V_{ds}(1+\lambda V_{ds})}{[1+(V_{ds}/V_{sate})^{m_{id}}]^{1/m_{id}}} \qquad [1]$$

Here $V_{sate} = \frac{I_{sat}}{g_{ch}}$: is the effective extrinsic saturation voltage, $g_{ch} = \frac{g_{chi}}{1+g_{chi}(R_s+R_d)}$: the extrinsic channel conductance, $g_{chi} = \frac{qn_s\mu}{L}$ is the intrinsic channel conductance, V_{ds} is the extrinsic drain-to-source voltage, m_{id} is a fitting exponent for the knee shape in the output characteristics, λ is an empirical factor for channel length modulation, R_s, R_d are the source and drain contact resistances, q is the elementary charge. The channel carrier density is

$$n_s = \frac{n_{s1}}{1+\left(\frac{n_{s1}}{n_{max}}\right)^{m_{ns}}} \qquad [2]$$

Here $n_{s1} = n_0\,ln\left[1 + exp\left(\frac{V_{gt}}{\eta V_{th}}\right)\right]$; $n_0 = C_{1d} \cdot \frac{V_{th}}{q}$; $C_{1d} = \frac{C_Q C_{1g}}{C_Q+C_{1g}}$; $C_{1g} = \frac{2\pi\varepsilon_0\varepsilon_r}{Arccosh\left[\frac{2d_i}{d_{cnt}}+1\right]}$; n_{max} is the maximum sheet density, m_{ns} is the transition parameter; $V_{gt} = V_{gs} - V_T$: the gate voltage swing; V_T: the threshold voltage; V_{th}: the thermal voltage; n_0 is the electron density at threshold; η: the sub-threshold ideality factor; C_{1d}: the gate to channel capacitance per unit length C_{1g}: the gate insulator capacitance d_i: the gate insulator thickness; d_{CNT}: the CNT diameter, C_Q: the quantum capacitance. The saturation current

$$I_{sat} = \frac{g_{chi}V_{gte}}{1+g_{chi}R_s+\sqrt{1+2g_{chi}R_s+\left(\frac{V_{gte}}{V_L}\right)^2}}. \qquad [3]$$

Here $V_L = \frac{v_s L}{\mu}$; $\mu = \frac{\mu_{fet1}}{1+(\theta V_{gte})^{m_t}}$, $\mu_{fet1} = \frac{1}{\frac{1}{\mu_0}+\frac{1}{\mu_1\left(\frac{V_{gte}}{\eta V_{th}}\right)^{m_{mu}}}}$ and

$$V_{gte} = \eta V_{th}\left[1 + \frac{V_{gt}}{2\eta V_{th}} + \sqrt{\delta^2 + \left(\frac{V_{gt}}{2\eta V_{th}} - 1\right)^2}\right] \qquad [4]$$

Here I_{sat} is the saturation current; V_L is the characteristic voltage of velocity saturation; v_s is the effective saturation velocity; L is the CNT length (gate length); μ is the channel mobility; μ_0 is the above threshold mobility; μ_1 is the adjustable mobility parameter; θ is the mobility degradation parameter; m_t is the mobility degradation exponent; m_{mu} is the adjustable mobility exponent; δ is the parameter controlling the transition width between above and below threshold. The capacitance-voltage characteristics are

$$C_{gs} = \frac{2}{3}C_{gc}\left[1 - \left(\frac{V_{sate}-V_{dse}}{2V_{sate}-V_{dse}}\right)^2\right] \qquad [5]$$

$$C_{gd} = \frac{2}{3}C_{gc}\left[1 - \left(\frac{V_{sate}}{2V_{sate}-V_{dse}}\right)^2\right] \qquad [6]$$

Here C_{gs} is the gate-to-source capacitance; C_{gd} is the gate-to-drain capacitance, $C_{gc} = C_{1d}L\,exp(V_{gt}/\eta V_{th})/[1 + exp(V_{gt}/\eta V_{th})]$ is the differential gate-to-channel capacitance, and V_{dse} is the effective and extrinsic drain-to-source bias.

Fig. 6 shows measured (symbols) and simulated (lines) CNTFET IVs (from [9]).

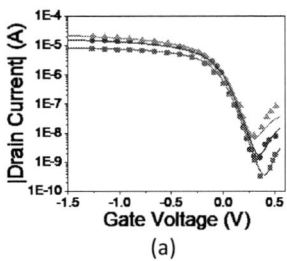 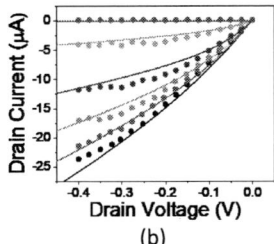

(a)   (b)

Fig. 6. Measured (symbols) and simulated (lines) CNTFET IVs. (a) The transfer characteristics at drain voltage, Vds = -0.1, -0.2, and -0.3 V, (b) the output characteristics at the gate voltage, Vgs from -1.3 to 0.2 V in 0.3 steps. (From [9]) ©IEEE2020

This model has been applied for the simulation of the CNTFET THz response and for the analysis of the CNTFET operation as a THz spectrometer. Fig. 7 (from [10] shows an example of the THz detection using CNTs. As seen, the detection is noisy and broadband but it could be improved using higher quality and more uniform CNTs.

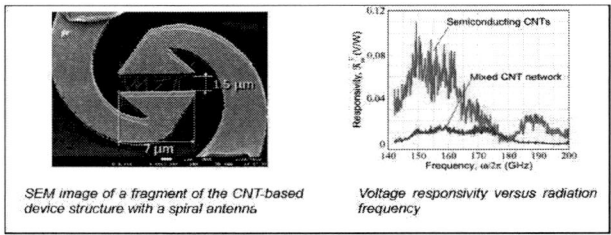

Fig. 7. THz detection using CNTs. (From [10])

Fig. 8 and 9 show the schematic of the simulation circuit for the CNT FET THz response and the schematic of the compact model used for the spectrometer simulation

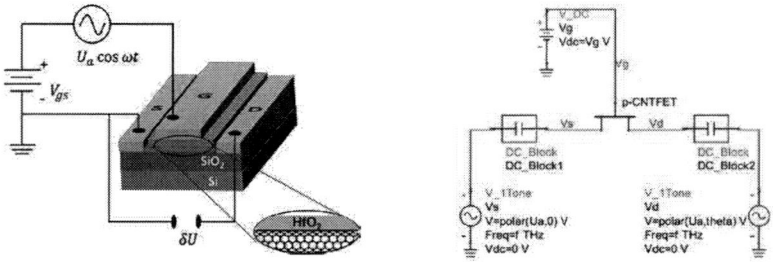

Fig. 8. Schematic of the simulation circuit for the CNT FET THz response

Fig. 9. Compact model used for the spectrometer simulation. [9]) ©IEEE2020

Fig. 10 and Fig. 11 show the simulated CNT FET THz spectrometer response.

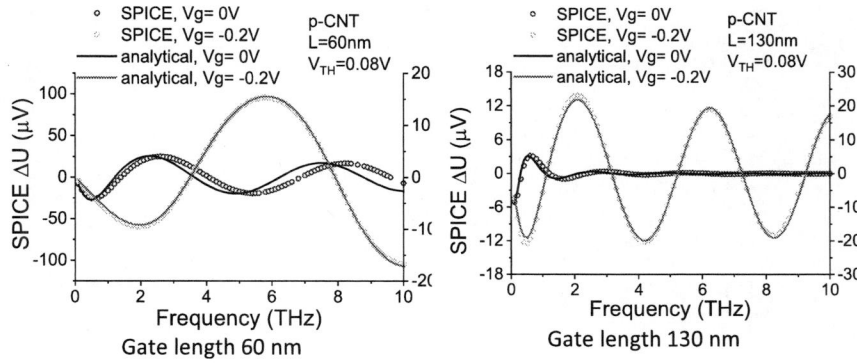

Fig. 10. CNT FET THz spectrometer response. (From [9]) ©IEEE2020

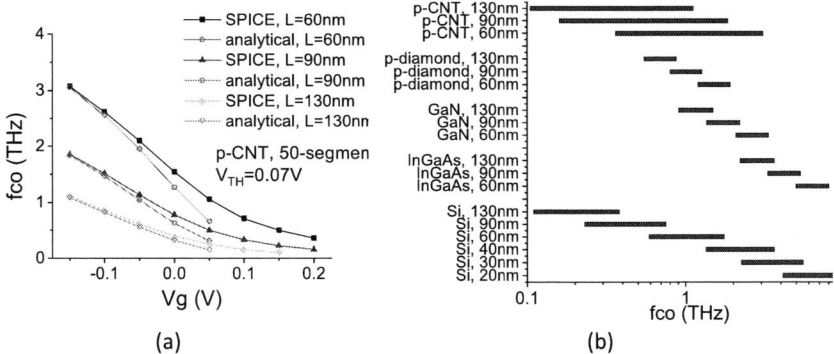

Fig. 11. CNT FET THz spectrometer measured frequency as a function of the gate bias (a) and the comparison of CNT FET THz spectrometer operating range with that of other materials systems. (From [9]) ©IEEE2020

As seen from Fig. 11 (a), the multi segment compact THz SPICE model [11] was used for the CNT FET THz spectrometer simulations

CNT mats and Thin Film Transistors

Fig. 12 presents CNT random and organized arrays. As mentioned above, CNT random mats have been used for the improved flexible interconnects. [1] Another application is using CNT mats as a sensing element [12], see Fig. 13. And CNT mats have been used for Thin Film Transistors (see Fig. 14.)

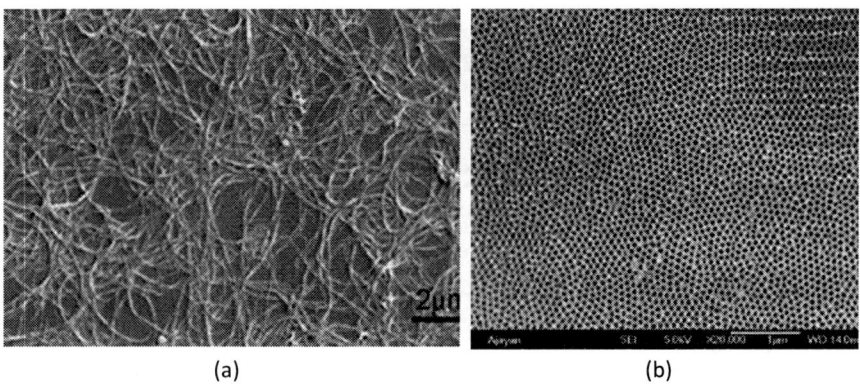

(a)   (b)

Fig. 12. Typical CNT array (a) and SEM micrograph of self-organized CNTs grown in aluminum anodic oxide films (From [13])

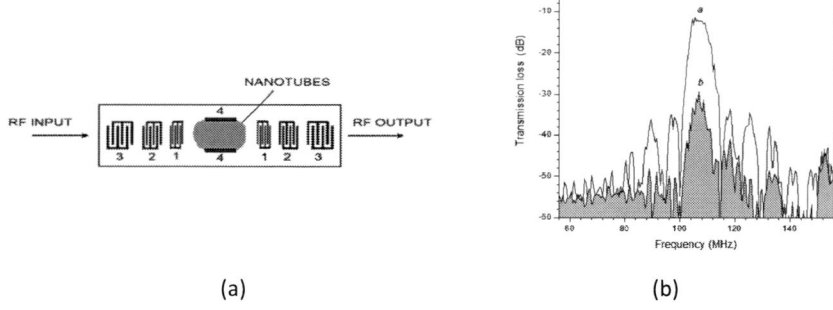

(a)   (b)

Fig. 13. Effect of Carbon Nanotubes on SAW propagation: (a) measurement setup; (b) transmission comparison. (From [12])

Fig. 14. CNT Thin Film Transistor.

The CNT sensing capabilities, their applications for flexible electronic systems and flexible interconnects and having CNT TFT capabilities make CNT mats uniquely suitable for applications in sensitive skin for robots (see Fig. 15 and reference [14])

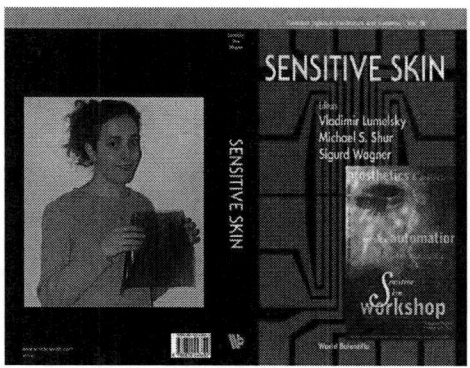

Fig. 15. Sensitive skin proceedings of DARPA/NSF workshop [14].

The CNT mat resistance diverges near the percolation point (see Fig. 16) and, near the percolation point, the CNT mat responsivity to external stimuli is diverging.

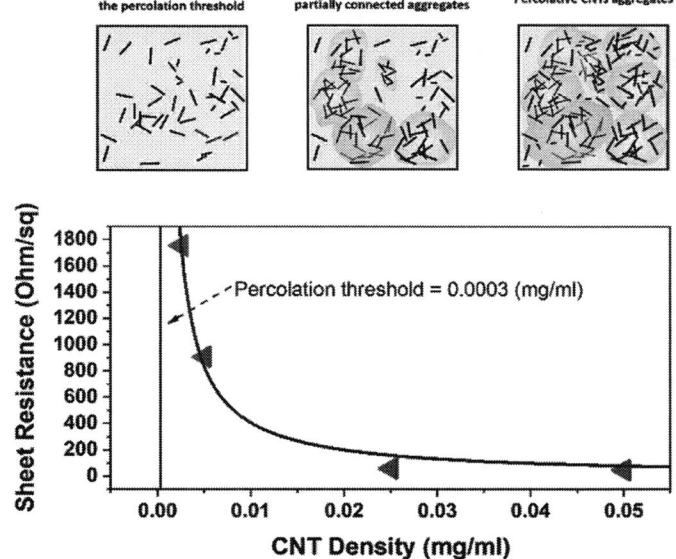

Fig. 16. Sheet resistance of CNT random networks as a function of the CNT density (the data from [2]).

Our THz transmission measurements showed that percolation point is shifted by a THz radiation rectified by asymmetric CNT contacts in the CNT mat [2], see Fig. 17.

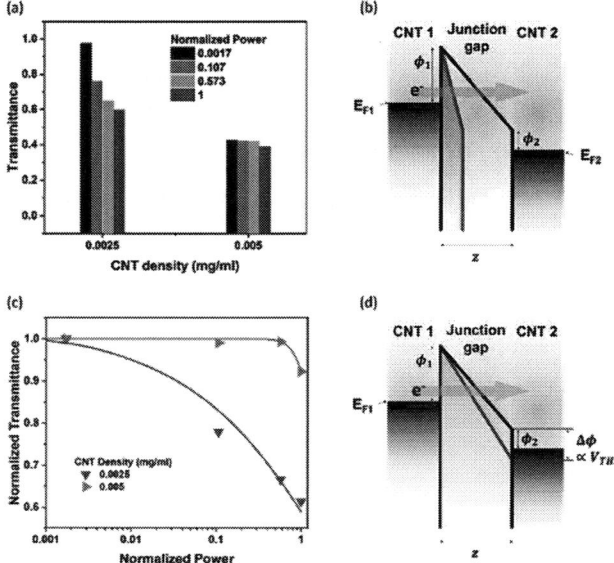

Fig. 17. Transmittance of CNT mats. at different incident beam powers, (b) energy band-diagram of the CNT-CNT tunneling junction; (c) Normalized transmittance against normalized incident beam power [2] d) Simplified energy band-diagram of the CNT-CNT tunneling junction for illustrating the reduction in transmittance with increasing the beam power due to the barrier height modulation.

Near the percolation point, the barrier between the contacting CNTs in the random CNT mat network is affected by the gate bias. As a result, the threshold voltage changes (see Fig. 18) decreasing the subthreshold slope as shown in Fig. 19.

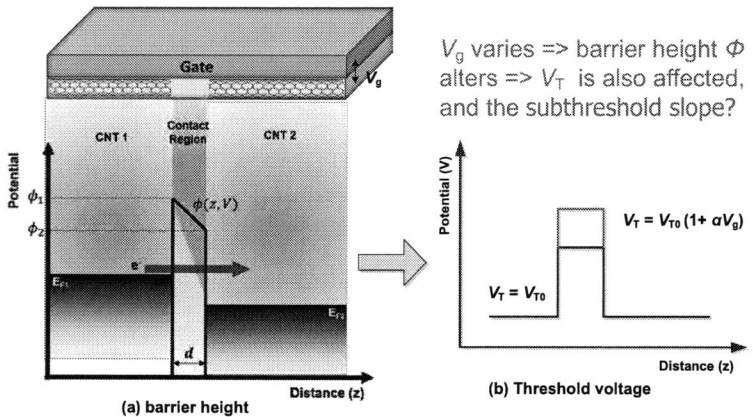

Fig. 18. Mechanism of the changing the subthreshold slope near the percolation point.

We introduced two empirical models to simulate the effect of the subthreshold slope reduction (see Fig. 19 and Fig. 20):

$$V_T = V_{T0} \, (1 + \alpha V_g)$$ [7]

$$V_T = V_{T0} - \beta (V_g - V_{T0})$$ [8]

The empirical parameters α and β depend on how close the CNT mat forming the TFT channel to the percolation point.

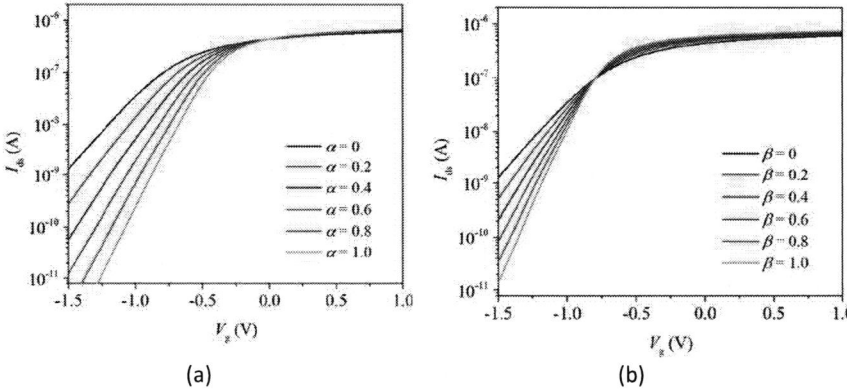

(a)   (b)

Fig. 19. Simulated CNT TFT current-voltage characteristics.

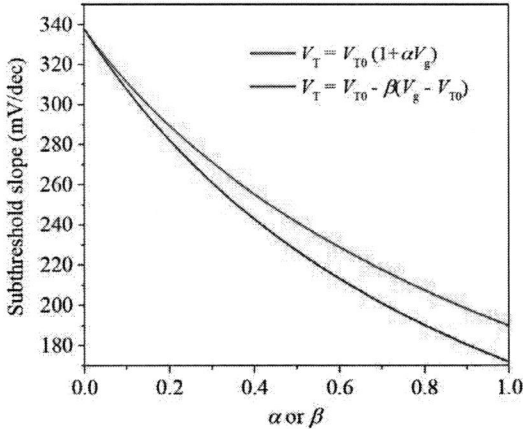

Fig. 20. Simulated subthreshold slope reduction.

As seen, our simulations predict a dramatic reduction of the subthreshold slope by using the CNT mats in the CNT TFT channel to be close to the percolation point.

Conclusions

In conclusion, we considered emerging applications of the CNT technology and described a compact multi segment model used to predict the performance of the CNT transistors up to the THz frequency range. In a short run, the most promising applications are for the CNT mats to be used in CNT TFTs, sensors, and flexible interconnects. Especially important are applications of such CNT mat near the percolation point, where their responsivity to external stimuli is diverging. Our THz transmission measurements showed that percolation point is shifted by a THz radiation rectified by asymmetric CNT contacts in the CNT mat. We now propose using the shift of the percolation point by the gate bias for reducing the CNT TFT sub threshold slope. Our calculations show that the reduction could be nearly a factor of two.

Acknowledgments

This work at RPI was supported by the U.S. Army Research Laboratory under the Cooperative Research Agreement (Project Monitor Dr. Meredith Reed), by the US ONR (Project Monitor Dr. Paul Maki), and by AFOSR (Project Monitor Dr. Ken Goretta).Place acknowledgments at the end of the text, before the references.

References

1. J. Park, X. Liu , T. Ytterdal , Jin Ho Kim , J. Xu, and M. Shur Carbon Nanotube Metal Polymer Composites for Flexible Active Interconnects, 2020 ECS Trans. 97, No , p. 383 (2020)
2. J. Park, Jin Ho Kim, J. Xu, and M. Shur, J. Park, J. H. Kim, J. Xu, and M. Shur, Terahertz transmission limiters using randomly oriented carbon nanotubes network near percolation threshold, ECS Trans. 97, No. 7 p. 329 (2020)
3. S. O. Koswatta, A. Valdes-Garcia, M. B. Steiner, Y.-M. Lin, and P. Avouris, "Ultimate RF performance potential of carbon electronics," IEEE Trans. Microw. Theory Techn. Vol. 59, no. 10, pp. 2739–2750, Oct. 2011.
4. J. Guo, S. Datta, and M. Lundstrom, M. Brink, P. McEuen, A. Javey, H. Dai, H. Kim, and P. McIntyre, Assessment of Silicon MOS and Carbon Nanotube FET Performance Limits Using a General Theory of Ballistic Transistors, IEDM Technical Digest. Pp. 711-714 (2002)
5. G. Hills, C. Lau, A. Wright, S. Fuller, M. D. Bishop, T. Srimani, P. Kanhaiya, R. Ho, A. Amer, Y. Stein, D. M. Arvind, A. Chandrakasan and M. M. Shulaker, "Modern microprocessor built from complementary carbon nanotube transistors," Nature, 572, pp. 595–602 (2019).
6. D.-M. Sun, M. Y. Timmermans, Y. Tian, A. G. Nasibulin, E. I. Kauppinen, S. Kishimoto, T. Mizutani and Y. Ohno, "Flexible high-performance carbon nanotube integrated circuits," Nat. Nanotechnol., 6, pp. 156–161 (2011).

7. M. S. Shur, J. S. Dordick, and P. M. Ajayan, Fluid Delivery to Cells and Sensing Properties of Cells Using Nanotubes, Patent Publication Application, US2004/0186459 A1, Sep. 23 (2004)
8. S. Kumar, G., Singh, A. and Raj, B. (2019). Design and analysis of a gate-all-around CNTFET-based SRAM cell, J. Comput. Electron. (2017)
9. J. Park, X. Liu X, T. Ytterdal, and M. Shur, "Carbon Nanotube Detectors and Spectrometers for the Terahertz Range," Crystals, 10(7), 601, (2020)
10. V. Ryzhii, T. Otsuji, M. Ryzhii, V. G. Leiman, G. Fedorov, G. N. Goltzman, I. A. Gayduchenko, N. Titova, D. Coquillat, D. But, W. Knap, V. Mitin, and M. S. Shur, "Two-dimensional plasmons in lateral carbon nanotube network structures and their effect on the terahertz radiation detection", J. Appl. Phys. 120, 4, 2016.
11. X. Liu, T. Ytterdal, and M. Shur, Plasmonic FET Terahertz Spectrometer, IEEE Access, pp. 1-6, 2020. DOI: 10.1109/ACCESS.2020.2982275; also arXiv:2001.06101v1 [physics.app-ph] 16 Jan 2020
12. D. Ciplys, S. Rumyantsev, M.S. Shur, R. Vajtai, B. Wei, P. Ajayan, R. Gaska, and R. Rimeika, Attenuation of Surface Acoustic Waves by Carbon Nanotubes, in MRS Proceedings Vol. 750, Y5.46, MRS Fall 2002, Symposium Y, Surface Engineering 2002--Synthesis, Characterization and Applications
13. T. Borca-Tasciuc, R. Vajtai, B. Q. Wei, and P. M. Ajayan, M. S. Shur, J. Deng and R. Gaska in Thermal Challenges in Next Generation Electronic Systems, Joshi & Garimella (eds.), Millpress, Rotterdam, ISBN 90-77017-03-8 (2002), pp. 79-84
14. V. Lumelsky, M. S. Shur, and S. Wagner, Editors, Sensitive Skin, World Scientific, ISBN 981-02-4369-3, 2000. Also published as Special Issue, International Journal of High Speed Electronics and Systems. International Journal of High Speed Electronics and Systems. Vol. 10, No 2, 2000

Optimizing Material Systems for All-Inkjet-Printed Organic Thin-Film Transistors

C. Jiang[a,b], and A. Nathan[c]

[a] Department of Engineering, Electrical Division, University of Cambridge, Cambridge CB3 0FA, UK
[b] Department of Clinical Neurosciences, Clifford Allbutt Building, University of Cambridge, Cambridge CB2 0HA, UK
[c] Cambridge Touch Technologies, 154 Cambridge Science Park, Cambridge CB4 0GA, UK

> All-inkjet-printed organic thin-film transistors enjoy advantages of low-cost printability and mechanical flexibility, thereby enabling newly emerging application areas. Recent developments have shown devices with low operating voltages and steep subthreshold slopes, using a bottom-gate bottom-contact structure. However, it would be also interesting to understand the optimization of the underlying materials system. This study investigates the composition and associated solvents for semiconductor inks, as well as the necessity of encapsulation. Basically, we compare the semiconductor inks of 6,13-bis(triisopropylsilylethynyl)pentacene with and without polystyrene binder, and find the importance of the polymer binder in lowering trap state density. Comparing semiconductor inks of different boiling points, e.g. toluene (low) and anisole (high), suggests that using a solvent with a high boiling point can enhance semiconductor crystallization. Using encapsulation with a fluoropolymer CYTOP is essential to reduce the trap state density. These results are important for further development of novel all-inkjet-printed organic thin-film transistors.

Introduction

With recent developments in organic electronic materials and device structures, organic thin-film transistors (OTFTs) have demonstrated novel applications, such as e-skin (1), flexible display (2), and wearables (3). These can be attributed to the advantages of OTFTs, including mechanical flexibility and low-cost printability (4). However, printed organic transistors typically require relatively high operating voltages, as compared to conventional vacuum-processed counterparts (5). This could be attributed to the higher defect density in the printed thin films (6), thus compromising steep subthreshold behavior (7–9). By reducing the semiconductor-dielectric interface density, an all-inkjet-printed OTFTs with a low operating voltage of 3 V had been reported in previous work, where a bottom-gate bottom-contact structure was used (10). Here, we used a small-molecule organic semiconductor with a polymer binder and a fluoropolymer encapsulation layer. However, the effectiveness of the materials selection needs to be further investigated.

In this work, we examine the materials used in our all-inkjet-printed OTFTs, in particular, addressing the polymer binder used in semiconductor ink, the encapsulation layer, and the solvent for semiconductor ink. We compare device performance of the all-inkjet-printed devices with and without the polymer binder, with and without the encapsulation layer, and using different solvents. Based on this, we highlight the usefulness of the polymer binder in the semiconductor, the fluoropolymer in protecting the devices from defects from the ambient environment, and the effect of the high boiling point solvent on semiconductor crystallinity. We believe that the results presented in this work and reference (11) should provide sufficient insight for future development of printed electronics with a broader range of materials system or device structures.

Experimental

Materials and ink formulations

Silver (Ag) was used for conductive electrodes, poly(4-vinylphenol) (PVP) was used as gate dielectric layer, 6,13-bis(triisopropylsilylethynyl)pentacene (TIPS-pentacene) was used as the semiconducting layer, and CYTOP was used for encapsulation. The Ag ink (jet-600C) was purchased from Hisense Electronics, Kunshan, China. The CYTOP (CTL-809M) and its solvent (CT-Solv. 180) were provided by Asahi Glass. The rest of chemicals were purchased from Sigma-Aldrich.

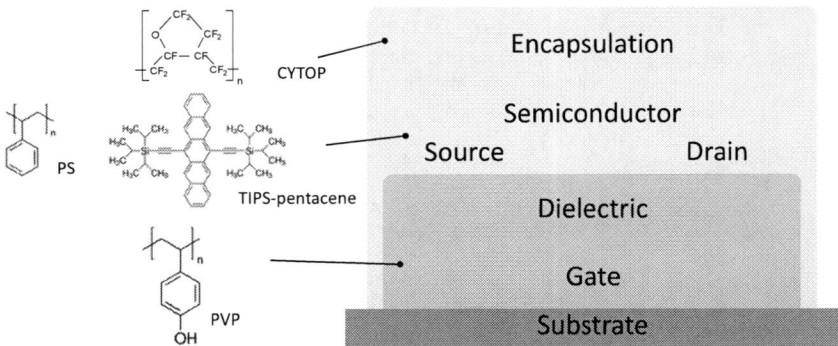

Figure 1. Materials used and cross-section of the device structure.

The gate dielectric ink was prepared by dissolving PVP in propylene glycol monomethyl ether acetate (PGMEA) at a concentration of 80 mg/mL, with the cross-linking reagent poly(melamine-co-formaldehyde) (PMF) mixed into the solution at a mass ratio of 1:2 to PVP. For the semiconductor inks, we compared TIPS-pentacene with and without PS. TIPS-pentacene and PS were dissolved in toluene at the same concentration of 10 mg/mL. The TIPS-pentacene solution was used as the ink of TIPS-pentacene without PS, whereas the ink of TIPS-pentacene with PS was prepared by mixing the two solutions at a volume-to-volume ratio of 3:1 (TIPS-pentacene:PS). For the semiconductor inks using anisole, TIPS-pentacene and PS were dissolved in anisole at the concentration of 10 mg/mL, and the two solutions were mixed at a volume-to-volume ratio of 3:1 (TIPS-pentacene:PS). The encapsulation ink was formulated by diluting the supplied CYTOP with its solvent at a weight-to-weight ratio of 3:1.

Device fabrication

The bottom-gate bottom-contact device structure was used in this study (as shown in Figure 1), to enhance device performance. All the inkjet printing processes were conducted in ambient air with a Dimatix material inkjet printer, DMP-2831. All the inks were filled into DMPLCP-11610 cartridges (Dimatix), of which the nozzles typically jet droplets at the size of around 10 pL. The waveform setting for nozzles was the same as the reported setting in the previous study (10).

The Ag ink was printed on a polyethylene naphthalate (PEN) (Teonex®, Dupont) substrate at a drop spacing of 50 μm and then annealed at 150 °C for 15 min to form conductive gate electrodes. Then, the dielectric ink was printed at a drop spacing of 5 μm. The printed dielectric film was baked at was 150 °C for 1 hour. Subsequently, the Ag ink was printed again and then annealed at 130 °C for 15 min to form source/drain electrodes, which were treated by a PFBT-ethanol solution (PFBT:ethanol = 1:1000) for 3 min and rinsed with ethanol. The semiconductor ink was printed at a drop spacing of 5 μm, followed by annealing at 100 °C for 30 min. Finally, CYTOP ink was printed and annealed at 100 °C for 30 min before device measurements.

Parameters extraction

To compare the device performance of devices, several parameters were extracted from measured device transfer (I_D/V_{GS}) characteristics, including mobility (μ), threshold voltage (V_{th}), on/off ratio (I_{on}/I_{off}), subthreshold slope (SS) and trap density (N_{SS}). The mobility and threshold voltage were calculated by fitting the square root of I_D versus V_{GS} using the following equation:

$$I_D = \frac{\mu C_i W}{2L}(V_{GS} - V_{th})^2 \qquad [1]$$

where L is the channel length, W the channel width and C_i the gate dielectric capacitance per unit area. The N_{SS} can be estimated using the following equation (12):

$$N_{SS} = \frac{C_i}{q^2}\left(\frac{SS}{\ln(10)}\frac{q}{k_B T} - 1\right) \qquad [2]$$

where k_B is Boltzman's constant, T the absolute temperature, q the elementary charge.

Results and Discussion

Semiconductors with and without a polymer binder

As mentioned in our previous reports (10), a polymer binder has been widely used to reduce the traps in printed TIPS-pentacene thin films (13–15). However, a considerable amount of work has also not used polymer binders (16–20), and some fabricated devices have exhibited better performance than the work using polymer binders.

Here, in order to verify the effectiveness of using a polymer binder in our study, all-inkjet-printed OTFTs were fabricated using semiconductor inks with and without PS. The electrical transfer characteristics of the devices were measured. As shown in Figure 2, the device without PS exhibited a lower on-state current, a smaller on/off ratio, and a less steep subthreshold slope.

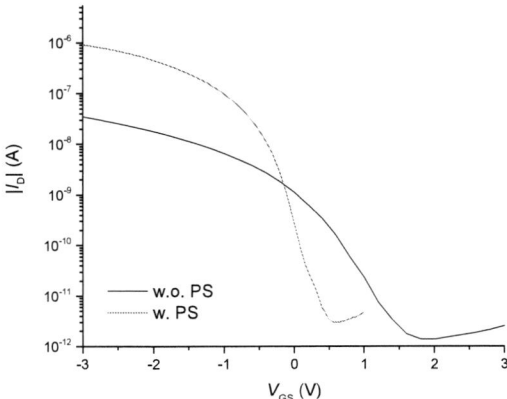

Figure 2. Measured electrical transfer (I_D–V_{GS}) characteristics of the fabricated OTFTs with and without PS.

The extracted parameters that compare the all-inkjet-printed OTFTs with the different conditions are listed in Table I. The most striking differences in the extracted parameters are the higher mobility and the lower N_{SS} in the all-inkjet-printed OTFTs with PS. These can be explained by the enhanced crystallinity and reduced traps in the printed TIPS-pentacene/PS thin films. Several reports demonstrated that a blend of small molecule organic semiconductor and a polymer binder was useful to reduce the N_{SS} (21–23), due to the better semiconductor crystallisation induced by vertical phase separation (24), which could be verified by transmission electron microscopy (TEM). For example, as reported by Niazi et al., the cross section TEM of the blade coated TIPS-pentacene/polymer (PαMS) blend film clearly illustrated the vertical phase separation of TIPS-pentacene and PαMS, in which PαMS was sandwiched by two thin layers of TIPS-pentacene (25). In principle, the phase separation that was associated with this work could be attributed to the different solubility parameters between the small molecule organic semiconductor and the polymer binder (15, 24). Since TIPS-pentacene with a smaller molecular weight (of 639.07) has larger solubility in toluene compared to PS (of which the molecular weight is as large as 35,000), phase separation between TIPS-pentacene and PS took place, resulting in better TIPS-pentacene crystallinity (24). As a result, the charge carrier trapping effect is less likely to occur in the printed TIPS-pentacene/PS thin films, compared to the printed TIPS-pentacene-only films. Therefore, the all-inkjet-printed OTFTs with PS demonstrate a higher mobility and a lower N_{SS}. It can be seen that by using PS as a polymer binder in the semiconductor layer, the traps in the all-inkjet-printed OTFTs can be effectively reduced.

TABLE I. Comparison of the all-inkjet-printed OTFTs with and without PS.

	μ (cm^2 V^{-1} s^{-1})	V_{th} (V)	I_{on}/I_{off}	SS (V/dec)	N_{SS} (eV^{-1} cm^{-2})
w.o. PS	0.004	0.59	10^4	0.476	1.48×10^{12}
w. PS	0.21	−0.13	10^5	0.123	2.24×10^{11}

Devices with and without encapsulation

Encapsulation is an important process for electronic devices based on organic materials, due to the oxidation and degradation of the organic materials by oxygen and moisture. For the all-inkjet-printed OTFTs in our previous report (10), CYTOP is used as the encapsulation layer. CYTOP is essentially a fluoropolymer, which is efficient in preventing moisture migration. The major substance in its solvent is perfluorotributylamine, which is also fluorinated. The advantages of using this fluorinated solvent are as follows: the surface energy of this fluorinated solvent is very low, so it can wet on most of materials; and, this fluorinated solvent does not dissolve most non-fluorinated materials. Since the semiconductor is TIPS-pentacene, a small molecule material that can be easily dissolved by most unsaturated organic solvents, using CYTOP to encapsulate TIPS-pentacene is a good solution for passivation.

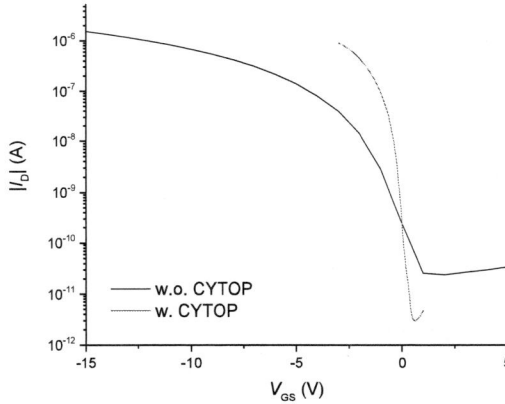

Figure 3. Measured electrical transfer (I_D–V_{GS}) characteristics of the fabricated OTFTs with and without PS.

Figure 3 depicts the electrical characteristics of the all-inkjet-printed OTFTS with and without the CYTOP encapsulation. The device without the CYTOP encapsulation demonstrate much higher operating voltage, with a gradual (not steep) subthreshold slope. In addition, the off-state current is also higher. This can be explained by the exposure of the back channel to the ambient environment, so the oxygen and moisture in the air interact with the back channel at the off state. The gradual subthreshold slope of the non-encapsulated device indicates the large number of traps in the device. As seen from the extracted parameters listed in Table II, the subthreshold slope of the non-encapsulated device is 1.1 V/dec, which is around one order higher than the encapsulated device (of which the subthreshold slope is 0.155 V/dec). In addition, the extracted trap density (N_{SS}) in the non-encapsulated device is also one order larger than the device with CYTOP encapsulation. The higher N_{SS} is the result of exposing the semiconductor layer to the ambient environment, which increases the possibility of degradation of the semiconductor and the creation of traps. Therefore, the non-encapsulated all-inkjet-printed OTFTs demonstrate poorer overall device performance, and the encapsulation layer of CYTOP is essential for the low-voltage all-inkjet-printed OTFTs in our previous report (10).

TABLE II. Comparison of the all-inkjet-printed OTFTs with and without CYTOP encapsulation.

	μ (cm^2 V^{-1} s^{-1})	V_{th} (V)	I_{on}/I_{off}	SS (V/dec)	N_{SS} (eV^{-1} cm^{-2})
w.o. CYTOP	0.04	−0.94	10^4	1.1	4.05×10^{12}
w. CYTOP	0.26	−0.17	10^5	0.155	3.70×10^{11}

Semiconductor ink using different solvents

The solvent used in the fabrication of all-inkjet-printed OTFTs described previously is toluene. Toluene (also known as methylbenzene) is an aromatic hydrocarbon, so it dissolves TIPS-pentacene well with a high saturation concentration of 6.57 wt. % at 23 °C, i.e., 61 mg/mL. Besides toluene, TIPS-pentacene can be dissolved in ketones and aromatic hydrocarbon, and the most widely used solvents are toluene, chlorobenzene and dichlorobenzenes. However, chlorinated-hydrocarbons are not compatible with the cartridges for Dimatix inkjet printers. Toluene is an aromatic solvent that exhibits good compatibility, so it was chosen in our previous work (10). However, the boiling point of toluene is relatively low (110 °C), so the drying rate of the semiconductor ink is too fast for good crystallisation of printed TIPS-pentacene thin films. Therefore, a solvent with a high boiling point (> 150 °C) and good compatibility with the cartridges is important.

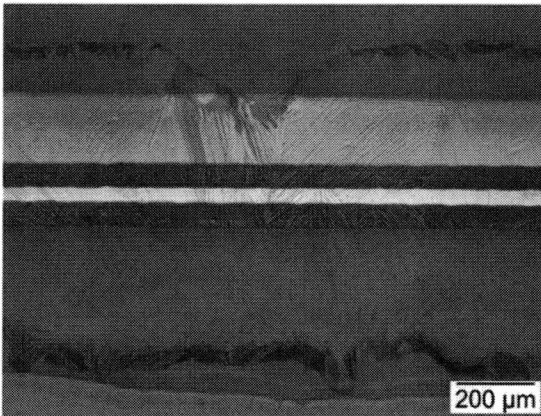

Figure 4. Polarized optical micrographs of all-inkjet-printed OTFTs using anisole as the solvent for the semiconductor ink.

The cartridges are compatible with a wide range of solvents. These include: aliphatic alcohols; aromatic hydrocarbons such as anisole and trimethylbenzene; aliphatic hydrocarbons such as hexane, dodecane; cellusolves; glycols; lactate esters; aliphatic and aromatic ketones including tetrahydrofuran (evaporates quickly); polyethylene glycols, and polypropylene glycols (11). Among aromatic solvents that possess compatibility with the cartridges, trimethylbenzenes have high boiling points of around 170 °C. However, one disadvantage of using a trimethylbenzene is the hazard of using it in the ambient environment. Aside from trimethylbenzenes, anisole is a good alternative. It has a relatively high boiling point of 155 °C and is not a hazardous substance. In addition, the solubility of TIPS-pentacene is 2.03 wt. %, which is adequate for formulating a semiconductor ink, typically around 1 wt. % or 10 mg/mL (11). Therefore, anisole is used as an alternative solvent for the semiconductor ink.

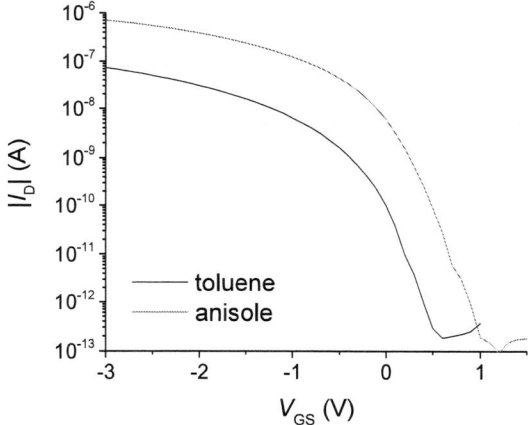

Figure 5. Measured electrical transfer (I_D–V_{GS}) characteristics of the fabricated OTFTs with and without PS.

As shown in Figure 4, the printed TIPS-pentacene thin films demonstrate a needle-like crystal structure in the channel area. These needle-like crystals are similar to those obtained by the well-controlled blade coatings (25, 26), and are vital to achieve high mobility of the fabricated TIPS-pentacene OTFTs.

TABLE III. Comparison of the all-inkjet-printed OTFTs using the anisole-based and toluene-based semiconductor inks.

	μ (cm^2 V^{-1} s^{-1})	V_{th} (V)	I_{on}/I_{off}	SS (V/dec)	N_{SS} (eV^{-1} cm^{-2})
Anisole-based	1.06	0.25	10^6	0.193	1.39×10^{11}
Toluene-based	0.11	−0.11	10^5	0.191	1.36×10^{11}

The electrical transfer characteristics of an all-inkjet-printed OTFT using the anisole-based semiconductor ink were measured, and, in comparison, a device using the toluene-based ink was measured as the control (see Figure 5). The anisole-based device demonstrates a one-magnitude higher on-state current than the toluene-based device. The extracted mobility, shown in Table III, indicates that the all-inkjet-printed OTFTs using anisole as the solvent for the semiconductor ink exhibit a higher mobility of 1.06 cm^{2}V^{-1}s^{-1}. This can be explained by the better crystallinity of the TIPS-pentacene thin films produced by the anisole-based ink. The higher on-state current provides a higher on/off ratio of 5.8×10^6 for the anisole-based OTFTs. Except for the mobility and on/off ratio, the all-inkjet-printed OTFTs using anisole-based and toluene-based inks demonstrate a similar subthreshold slope of 193 mV/dec and 191 mV/dec, respectively, and therefore a similar trap density of 1.39×10^{11} and 1.36×10^{11} eV^{-1}cm^{-2}, respectively.

Conclusions

This work examined the materials system used in our previously reported all-inkjet-printed OTFTs. The results show that it is essential to add a polymer binder to TIPS-pentacene ink to enhance device mobility and reduce trap density, thereby lowering device operating voltage. In addition, CYTOP is also crucial to significantly reduce trap density. Finally, the use of anisole, which has a high boiling point could help semiconductor crystallization during printing, thereby significantly improving the mobility of the printed OTFTs.

Acknowledgement

This work was supported by China Scholarship Council, Cambridge International Scholarship Scheme, Great Britain-China Educational Trust, Queens' College, EPSRC under project EP/M013650/1 and the EU under projects DOMINO 645760, 1D-NEON 685758, and BET-EU 692373.

References

1. J. Xu, S. Wang, G.N. Wang, C. Zhu, S. Luo, L. Jin, X. Gu, S. Chen, V.R. Feig, J.W.F. To, S. Rondeau-gagné, J. Park, B.C. Schroeder, C. Lu, J.Y. Oh, Y. Wang, Y. Kim, H. Yan, R. Sinclair, D. Zhou, G. Xue, B. Murmann, C. Linder, W. Cai, J.B. Tok, J.W. Chung, and Z. Bao, *Science*, **355**, 59 (2017).
2. M. Noda, N. Kobayashi, M. Katsuhara, A. Yumoto, S. Ushikura, R. Yasuda, N. Hirai, G. Yukawa, I. Yagi, K. Nomoto, and T. Urabe, *J. Soc. Inf. Disp.*, **19**, 316 (2011).
3. C. Jiang, H.W. Choi, X. Cheng, H. Ma, D. Hasko, and A. Nathan, *Science*, **363**, 719 (2019).
4. X. Guo, Y. Xu, S. Ogier, T.N. Ng, M. Caironi, A. Perinot, L. Li, J. Zhao, W. Tang, R.A. Sporea, A. Nejim, J. Carrabina, P. Cain, and F. Yan, *IEEE Trans. Electron Devices*, **64**, 1906 (2017).
5. C. Jiang, C. Xiang, and A. Nathan, *Proc. IEEE*, **107**, 2084 (2019).
6. E. Sowade, E. Ramon, K.Y. Mitra, C. Martínez-Domingo, M. Pedró, J. Pallarès, F. Loffredo, F. Villani, H.L. Gomes, L. Terés, and R.R. Baumann, *Sci. Rep.*, **6**, 33490 (2016).

7. M. Singh, H.M. Haverinen, P. Dhagat, and G.E. Jabbour, *Adv. Mater.*, **22**, 673 (2010).
8. J. Chang, X. Zhang, T. Ge, and J. Zhou, *Org. Electron.*, **15**, 701 (2014).
9. H.Y. Tseng and V. Subramanian, *Org. Electron.*, **12**, 249 (2011).
10. L. Feng, C. Jiang, H. Ma, X. Guo, and A. Nathan, Org. Electron. **38**, 186 (2016).
11. C. Jiang, All-Inkjet-Printed Low-Voltage Organic Thin-Film Transistors, 2018.
12. W.L. Kalb and B. Batlogg, *Phys. Rev. B*, **81**, 035327 (2010).
13. B.K.C. Kjellander, W.T.T. Smaal, K. Myny, J. Genoe, W. Dehaene, P. Heremans, and G.H. Gelinck, *Org. Electron.*, **14**, 768 (2013).
14. B.K.C. Kjellander, W.T.T.T. Smaal, J.E. Anthony, and G.H. Gelinck, *Adv. Mater.*, **22**, 4612 (2010).
15. S.Y. Cho, J.M. Ko, J. Lim, J.Y. Lee, and C. Lee, *J. Mater. Chem. C*, **1**, 914 (2013).
16. Y.H. Kim, B. Yoo, J.E. Anthony, and S.K. Park, *Adv. Mater.*, **24**, 497 (2012).
17. S. Chung, S.O. Kim, S.K. Kwon, C. Lee, and Y. Hong, *IEEE Electron Device Lett.*, **32**, 1134 (2011).
18. M. Medina-Sánchez, C. Martínez-Domingo, E. Ramon, and A. Merkoçi, *Adv. Funct. Mater.*, **24**, 6291 (2014).
19. J.A. Lim, W.H. Lee, D. Kwak, and K. Cho, *Langmuir*, **25**, 5404 (2009).
20. S.H. Lee, M.H. Choi, S.H. Han, D.J. Choo, J. Jang, and S.K. Kwon, *Org. Electron.*, **9**, 721 (2008).
21. L. Feng, W. Tang, X. Xu, Q. Cui, and X. Guo, *IEEE Electron Device Lett.*, **34**, 129 (2013).
22. W. Tang, L. Feng, C. Jiang, G. Yao, J. Zhao, Q. Cui, and X. Guo, *J. Mater. Chem. C*, **2**, 5553 (2014).
23. L. Feng, W. Tang, J. Zhao, Q. Cui, C. Jiang, and X. Guo, *IEEE Trans. Electron Devices*, **61**, 1175 (2014).
24. J. Kang, N. Shin, D.Y. Jang, V.M. Prabhu, and D.Y. Yoon, *J. Am. Chem. Soc.*, **130**, 12273 (2008).
25. M.R. Niazi, R. Li, M. Abdelsamie, K. Zhao, D.H. Anjum, M.M. Payne, J. Anthony, D.M. Smilgies, and A. Amassian, *Adv. Funct. Mater.*, **26**, 2371 (2016).
26. Y. Diao, B.C.K.K. Tee, G. Giri, J. Xu, D.H. Kim, H. a. Becerril, R.M. Stoltenberg, T.H. Lee, G. Xue, S.C.B.B. Mannsfeld, and Z. Bao, *Nat. Mater.*, **12**, 665 (2013).

Chapter 5

H03 – Applications in Displays, ICs, and Beyond

Display and LSI Applications of Oxide Semiconductor LSIs (OS LSIs) Using Crystalline In–Ga–Zn Oxide (IGZO): Applications Related to Coronavirus COVID-19 Pandemic

Tatsuya Onuki, Yuki Okamoto, Takeshi Aoki, Takanori Matsuzaki,Munehiro Kozuma, Hitoshi Kunitake, Ryosuke Motoyoshi, Hajime Kimura, Yasumasa Yamane, Shinya Sasagawa, and Shunpei Yamazaki

Semiconductor Energy Laboratory Co., Ltd., 398 Hase, Atsugi-shi, Kanagawa, 243-0036, Japan

> This paper shows structures and features of a field-effect transistor (FET) fabricated using a c-axis aligned crystalline In–Ga–Zn oxide (CAAC-IGZO), which is a typical example of crystalline oxide semiconductor (OS) ceramics, as a channel material, and applications of the CAAC-IGZO FET up to this point. In this study, we propose an AI accelerator as novel LSI using an OS (OS LSI) and make a detail description of the AI accelerator.

1 Introduction

The coronavirus COVID-19 pandemic is becoming a global crisis in 2020. Global climate change is an indirect cause of the spread of infectious diseases (1). The increase in greenhouse gas as a result of electric power generation to cover increased power consumption has an adverse effect on this climate change. The amount of power consumption continues to increase as human society advances, and Japan Science and Technology Agency's report says that global and Japan's annual power consumption concerning IT equipment in 2050 will be nearly 200 times as large as the present global and Japan's total annual power consumption." (2). This means that one of the main causes of the global climate change would be the increase in the amount of power consumption through the spread of fast-growing Internet of Things (IoT) and artificial intelligence (AI) technologies.

An answer to the question of how to reduce power consumption and curb the global climate change while developing the IoT and AI technologies is to reduce the power consumption of semiconductor devices and reduce the power consumption concerning IT equipment to 1/1000 of the present power consumption.

A field-effect transistor (FET) fabricated using a c-axis aligned crystalline In–Ga–Zn oxide (CAAC-IGZO), which is a typical example of crystalline oxide semiconductor (OS) ceramics, exhibits off-state current I_{off} in the order of yA/μm (10^{-24} A/μm) at 85°C (3,4). The CAAC-IGZO FET with such a feature would be one of the solutions to the increase in the amount of power consumption through the spread of the IoT and AI technologies.

Section 2 of this paper describes a CAAC-IGZO crystalline structure and a CAAC-IGZO FET device structure. Section 3 shows electrical characteristics and features of a

CAAC-IGZO FET. Section 4 introduces applications of the CAAC-IGZO FET up to this point. In Section 5, we have proposed an AI accelerator to reduce power consumption while developing the IoT and AI technologies. Section 6 is the summary of the study.

2 CAAC-IGZO

2.1. Topological Structure

CAAC-IGZO belongs to a new crystalline phase in a boundary region that is different from "Amorphous" and "Crystal" such as single crystal and polycrystal (5–10). The results of XRD analysis of the CAAC-IGZO crystalline structure and image analysis of the CAAC-IGZO crystalline structure through TEM images and Voronoi diagrams in the previous studies indicate that atoms are arranged in parallel in the c-axis direction like an $InGaZnO_4$ single crystal structure while neither clear alignment order nor a clear grain boundary is not observed in the a-b plane (7,11).

CAAC-IGZO is not influenced by grain boundary scattering; thus, when a CAAC-IGZO FET to be shown in Section 2.2 is fabricated, current that might be caused by a defect does not flow through a recombination center. That is probably the reason why the CAAC-IGZO FET exhibits an extremely low off-state current in the order of yA/μm (10^{-24} A/μm), which is the biggest feature to be described in Section 3. Moreover, CAAC-IGZO can be easily deposited by sputtering and is therefore suitable for mass production.

2.2. Device Structure of CAAC-IGZO FET

Figure 1(a) is a CAAC-IGZO FET schematic diagram. An FET which we have developed includes CAAC-IGZO as a channel material and is formed on an insulator. The CAAC-IGZO FET has a trench gate self-aligned (TGSA) structure and is a four-terminal element including a back gate BG (12,13). In the TGSA structure, a gate is formed in a self-aligned manner between a source (S) and a drain (D). This eliminates overlap between the gate and the S/D and thus the CAAC-IGZO FET has lower parasitic capacitance than an FET that has overlap between the gate and the S/D. In cutting-edge processes, CAAC-IGZO FETs with gate lengths of 25 nm, 21 nm, and 13 nm have been reported (13,14,15).

Figure 1(b) is a cross-sectional schematic diagram of a CAAC-IGZO FET with an encapsulation structure. In order to achieve stable threshold voltage and high reliability performance of the CAAC-IGZO FET, it is necessary to suppress generation of V_OH, which corresponds to a state in which a hydrogen atom H exists in an oxygen vacancy V_O, in CAAC-IGZO. We have adopted, in our manufacturing processes, the encapsulation structure of the CAAC-IGZO FET because it would suppress hydrogen diffusion from the outside into the FET and reduce hydrogen in the FET.

(a)

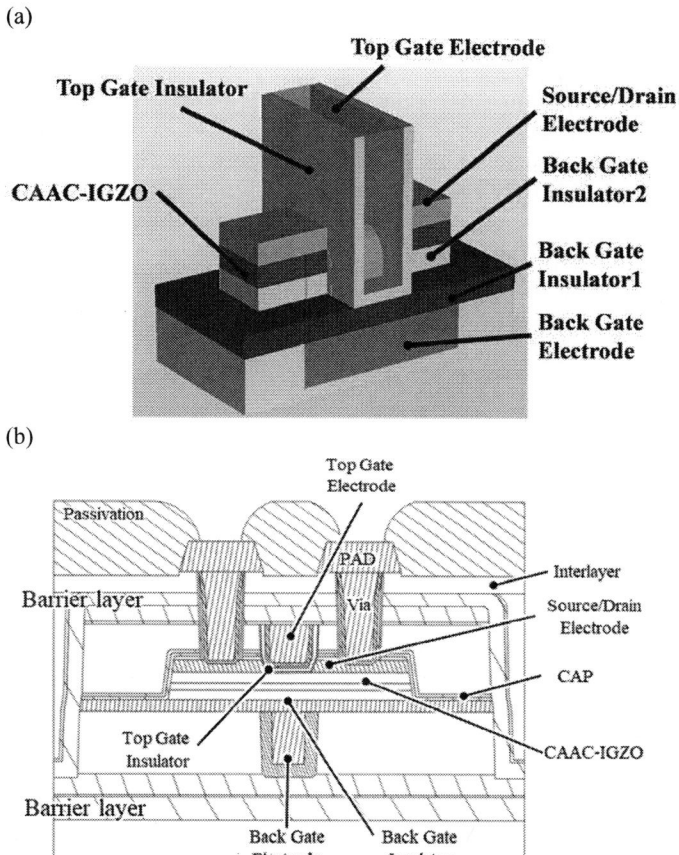

(b)

Figure 1. (a) A bird's view of the TGSA CAAC-IGZO FET and (b) a cross-sectional conceptual diagram of the CAAC-IGZO FET with the encapsulation structure.

3 Electrical Characteristics of CAAC-IGZO FET

This section shows electrical characteristics of CAAC-IGZO FETs fabricated in recent prototype lots. Electrical characteristics of TGSA CAAC-IGZO FETs with channel length L/channel width $W = 60$ nm/60 nm are shown in the subsections.

3.1 Static Characteristics

Static characteristics of the TGSA CAAC-IGZO FETs with $L/W = 60$ nm/60 nm are shown. Figure 2(a) shows drain current–gate voltage (I_d–V_g) curves when BG voltage V_{BG} is 0 V and drain voltage is 0.1 V or 1.2 V. One of the most remarkable aspects of the electrical characteristics of the CAAC-IGZO FET is that off-state current is lower than the measurement limit of measuring equipment (B1500A manufactured by Keysight Technologies). In addition, the field-effect mobilities of the CAAC-IGZO FETs have small temperature dependence and increase at high temperatures (13) (Fig. 2(b)).

Since each of the CAAC-IGZO FETs is a four-terminal element, the threshold voltage can be adjusted through control of the BG voltage (Fig. 2(c)). Consequently, the threshold voltage can be adjusted to be a desired value depending on an application.

(a)

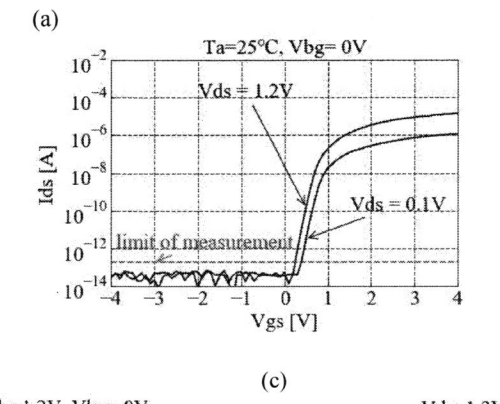

(b)   (c)

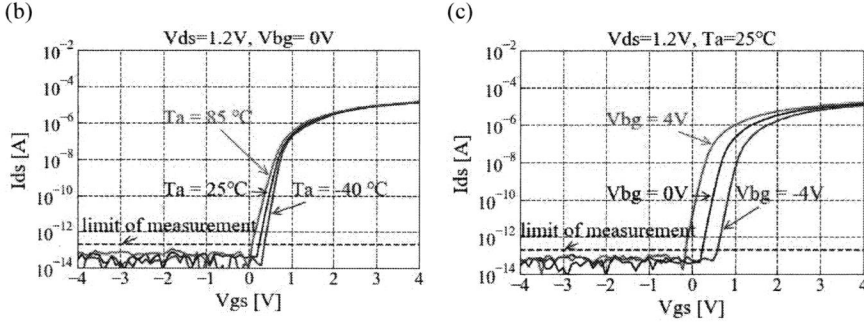

Figure 2. (a) I_d–V_g curves at $T_a = 25°C$ and $V_{bg} = 0$ V, (b) temperature dependence of I_d–V_g curves at $V_{bg} = 0$ V, and (c) BG voltage dependence of I_d–V_g curves at $T_a = 25°C$ of the CAAC-IGZO FETs with $L/W = 60$ nm/60 nm.

3.2. Off-state Current

The I_{off} of the CAAC-IGZO FET is as extremely low as below the measurement limit of measuring equipment and cannot be obtained by simply measuring the I_{off} of a single-device of the CAAC-IGZO FET. Accordingly, as described below, we have measured the I_{off} of the CAAC-IGZO FET (16). Figure 3(a) is a circuit diagram for I_{off} measurement. Even when a device under test (DUT) is off, a small amount of leakage current flows; therefore, the voltage of a floating node gradually decreases due to the leakage current and the voltage of an output OUT (V_{out}) decreases over time. We have estimated the I_{off} of the CAAC-IGZO FET from the decrease in V_{out} (ΔV). When V_{out} is monitored at regular time intervals as shown in Fig. 3(b), the estimated leakage current of the CAAC-IGZO FET with $L/W = 60$ nm/60 nm is 260 yA at 85°C, as shown in Fig. 3(c).

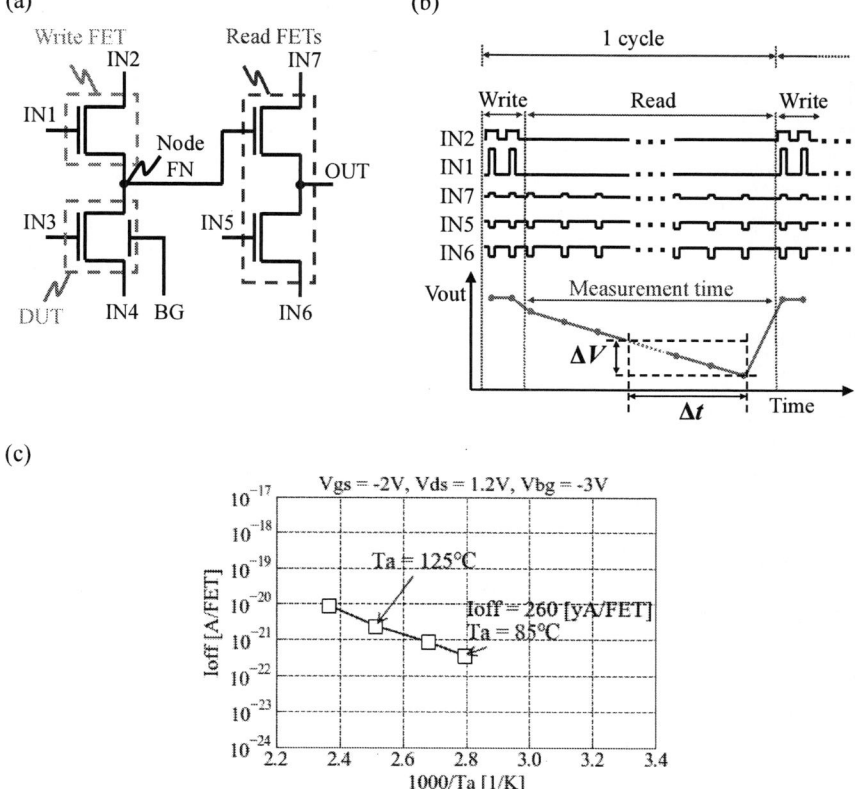

Figure 3. (a) A circuit diagram for I_{off} measurement, (b) a timing chart for I_{off} measurement, and (c) I_{off} measurement results.

3.3 Reliability Performance

We have evaluated the reliability performance of the fabricated CAAC-IGZO FETs (17). As reliability testing, we have performed positive gate bias-temperature stress testing (+GBT testing) under a stress voltage condition of an increase of 10% of operating voltage (3.3 V). The +GBT testing is performed to check characteristics variation of the CAAC-IGZO FETs over time by application of positive bias stress to gate electrodes. Specifically, we have performed +GBT testing at a temperature of 150°C by application of a stress of +3.63 V to the gate electrodes. Figure 4 shows +GBT testing results. The time during which the V_{sh} variation amount (ΔV_{sh}) is kept lower than 100 mV is defined as lifetime, and the estimated lifetime of one sample is longer than 6000 h, as can be seen from Fig. 4.

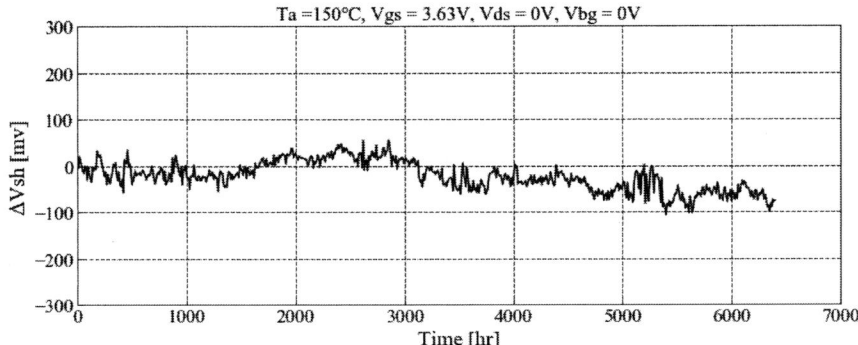

Figure 4. A ΔV_{sh}–time characteristics diagram showing +GBT testing results (stress conditions: 150°C and $V_g = 3.63$ V).

3.4. Cutoff Frequency f_T

We have evaluated the current characteristics of the CAAC-IGZO FETs in high-speed operation. The maximum driving force g_m for each gate load c_g is estimated from cutoff frequency. Cutoff frequency f_T is frequency when current gain becomes 1. The cutoff frequency f_T of a bulk Si FET with a gate length of 50 nm is approximately 205 GHz [18]. Figure 5 shows the measured cutoff frequency f_T of the CAAC-IGZO FETs with $L/W = 60$ nm/60 nm at $T_a = 25$°C. As shown in Fig. 5, the cutoff frequency f_T is 22.5 GHz. In addition, the threshold voltage of the CAAC-IGZO FETs is controlled by changing V_{bg} applied to the back gates. The change of V_{bg} may change load and mobility; however, the cutoff frequency f_T of the CAAC-IGZO FETs hardly change even when V_{bg} changes [17].

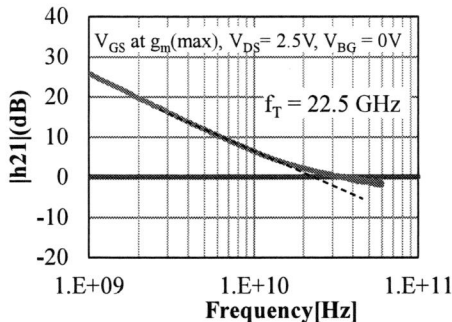

Figure 5. A current gain–cutoff frequency characteristics diagram.

3.5. Memory Endurance

In measurement of a memory cell circuit shown in Fig. 6(a), "H" and "L" are repeatedly written to a floating node FN. After a certain number of data write cycles, FN voltage after "H" writing and FN voltage after "L" writing are measured. The FN voltage is calculated by measurement of current I_{RBL} flowing to a terminal RBL and estimation of gate–source voltage V_{GS} from I_D–V_{GS} characteristics of read FETs.

Standard evaluation conditions are V_{WWL} = 3.3 V/0 V and V_{WBL} = 1.2 V/0 V. In addition, V_{WBG} is voltage at which the FET is reliably turned off at V_{WBL} = 0 V (e.g., V_{WBG} = −4 V), and T_a is 25°C. Figure 6(b) reveals that endurance is more than 10^{12} times. The previous study demonstrates endurance of more than 10^{14} times (17).

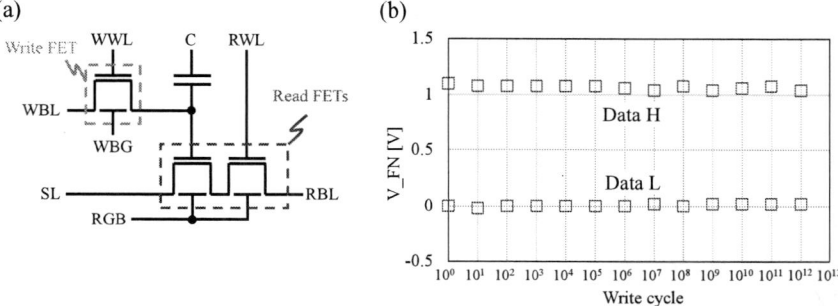

Figure 6. (a) A memory circuit diagram and (b) evaluation results of memory endurance.

3.6. 3D-Stacked CAAC-IGZO FET

As shown in Fig. 7, a CAAC-IGZO layer is formed in a CMOS wiring layer; thus, novel LSI using an OS (OS LSI) has a monolithically stacked structure of CMOS and CAAC-IGZO (12,19). In addition, as shown in Fig. 8(a), two or more layers of CAAC-IGZO FETs are monolithically stacked (20,21). When the BG voltage of the CAAC-IGZO FETs is independently controlled, the threshold voltage of the FETs is adjusted one by one. As can be seen from the I_d–V_g curves at drain voltage V_d = 1.2 V in Fig. 8(b), the first and second CAAC-IGZO FETs have subthreshold slopes (S.S.) of 91 and 100 mV/dec, saturation field-effect mobilities (μFE) of 10.3 and 4.9 cm^2/Vs, and on-state currents I_{on} of 5.5×10^{-6} and 2.2×10^{-6} A at V_g = 2.5 V. A vertical FET including In–Al–Zn–O (IAZO) in a channel layer has been reported (22). The monolithically stacked structure is one of the main features of the OS LSI. The use of the OS LSI enables LSI design with a high degree of freedom in a circuit configuration.

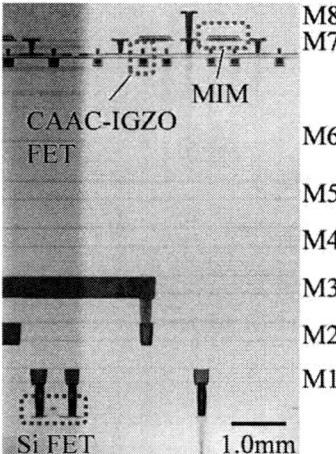

Figure 7. A cross-sectional view of a monolithically stacked structure of a Si FET and a CAAC-IGZO FET.

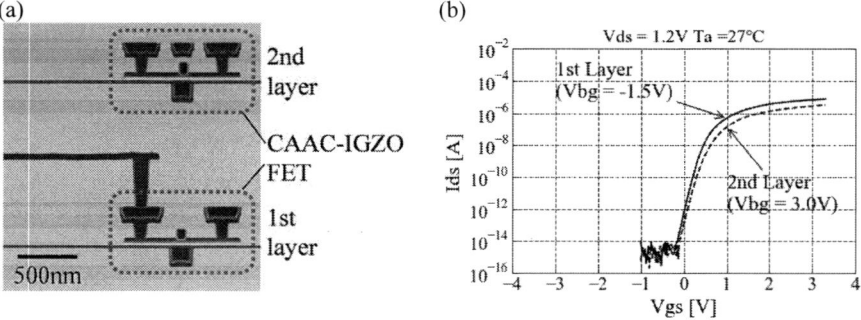

Figure 8. (a) A cross-sectional view and (b) I_d–V_g curves of a monolithically stacked structure of two layers of CAAC-IGZO FETs.

4 Application Using CAAC-IGZO

4.1. Application for Display

CAAC-IGZO has been widely used in displays. These displays are capable of performing idling stop driving that utilizes extremely low I_{off} of CAAC-IGZO FETs. The displays are used as low-power displays (23,24).

The number of people who do their work remotely has been increasing due to the COVID-19 pandemic, and the demand for displays for augmented reality (AR)/virtual reality (VR) would be increased. Our small display fabricated by CAAC-IGZO technology for AR/VR usage meets such demand (25). The fabricated display shown in Fig. 9 includes a CAAC-IGZO backplane and is an ultra-high resolution OLED display with a resolution over 5000 ppi.

(a)   (b)

Figure 9. (a) A photograph of the fabricated display and (b) an enlarged image of a display area.

4.2. Applications for LSI

4.2.1. OS Memory. In general, Si FETs are used in dynamic RAM (DRAM) memory cells. DRAM where Si FETs are replaced with CAAC-IGZO FETs is called dynamic oxide semiconductor RAM (DOSRAM) (26,27,28). DOSRAM achieves high operating frequency and long retention time.

Figure 10 shows a chip micrograph and a circuit diagram of a fabricated DOSRAM module. This module has a structure where an array of DOSRAM cells each composed of one CAAC-IGZO FET and one capacitor are monolithically stacked on a sense amplifier array composed of Si FETs. The DOSRAM module that uses CAAC-IGZO FETs with extremely low I_{off} is capable of retaining data stored in storage nodes for a long time. A normal DRAM cell needs refresh operation every several tens of milliseconds. In contrast, a DOSRAM cell only needs refresh operation less than once an hour in a wide temperature range from −40 to 125°C. Considering the entire system, the use of DOSRAM that has lower refresh cycles than DRAM leads to reductions in power consumption and operation standby time that accompanies refresh operation; therefore, the shorter operation standby time would result in better operation performance.

In the DOSRAM module, the bit line is extremely short because the DOSRAM cell is fabricated directly on the sense amplifier. As a result, the parasitic capacitance of the bit line is small, and high-speed read operation is possible. In addition, cell capacitance is reduced owing to low cell leakage, and high-speed write operation is also possible. The small cell capacitance enables fabrication of a capacitor with a lower aspect ratio than that of a trench capacitor used in the DRAM cell, which produces advantages of reductions in process cost and the degree of difficulty in processes.

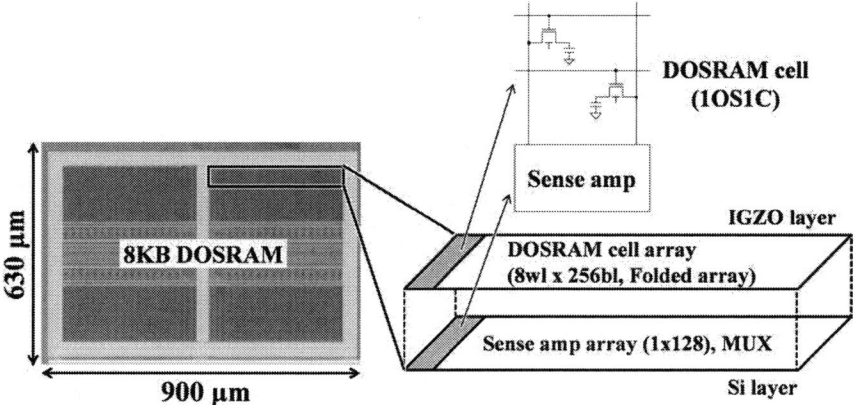

Figure 10. A chip micrograph and a circuit diagram of the fabricated DOSRAM module.

We have also reported nonvolatile oxide semiconductor RAM (NOSRAM) as another OS memory (26,29,30). Figure 11 is a NOSRAM cell circuit diagram. A NOSRAM cell is composed of two FETs and one capacitor. A CAAC-IGZO FET is used as a write FET M1 that requires extremely low I_{off}. As a read FET M2, we have used a CAAC-IGZO FET (29) or a Si FET (30), and an optimal circuit configuration varies depending on the application. In the case where all the FETs in the NOSRAM cell are CAAC-IGZO FETs, a sense amplifier for reading and a peripheral circuit are provide below the NOSRAM cells as in the DOSRAM module in Fig. 10.

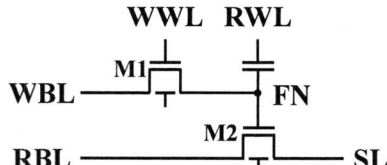

Figure 11. A NOSRAM cell circuit diagram.

4.2.2. NoffCPU. In an application where a sleep period is extremely long compared to an active period, such as an IoT sensor node, the proportion of static power during the sleep period to the total power consumption is high; therefore, what is called a normally-off (Noff) processor in which the supply of power supply voltage is usually stopped and power supply voltage is supplied only when needed has been proposed (31). Specifically, processors in which memory hierarchies such as a register and cache memory are made nonvolatile have been developed. Furthermore, processors to which MRAM (32) and STT-MRAM (33) are applied have been proposed.

The NOSRAM described in Subsection 4.2.1 would directly contribute to making cache memory nonvolatile. In addition, we have proposed backup and restore operations of register data in and from a nonvolatile block to which a NOSRAM or DOSRAM is applied by direct connection of the nonvolatile block to a register in a processor (26,27,34–38).

The use of a structure where a control circuit (nonvolatile block) for charge injection and retention of a floating node FN is added to a normal register (volatile block) as shown in Fig. 12 enables fabrication of what is called a nonvolatile register (26,27,34). The nonvolatile register is capable of performing data backup and restore operations in and from a local area; therefore, it eliminates the need for data exchange with external circuits of the processor and enables data backup and restore operations at low power consumption.

Furthermore, in an application where a sleep period is extremely long compared to an active period, operation for starting the supply of power supply voltage often accounts for much of the active period. Accordingly, achieving high-speed startup operation of a power supply circuit is also important.

In view of the above, we have proposed retention of reference voltage and reference current by using a structure where analog memory to which a NOSRAM or DOSRAM is applied to the inside of an integrated voltage regulator (IVR) as shown in Fig. 13 (34). We have fabricated a 32-bit microcontroller shown in Fig. 14 through a 110-nm Si FET/60-nm CAAC-IGZO FET hybrid process. When the supply of power supply voltage is restarted after one hour, processing is restarted from the state in which the supply of the power supply voltage is stopped. This indicates the possibility of fabrication of the Noff processor and shows the achievement of the high-speed startup operation of the power supply circuit. The microcontroller achieves a low power consumption of 880 nW in a standby period, a high-speed low-energy backup operation at 21 ns and 0.130 nJ, and high-speed system restore operation at 4.69 µs (34).

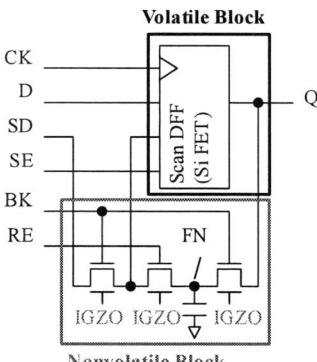

Figure 12. A circuit configuration example of a nonvolatile register using a normally-off processor.

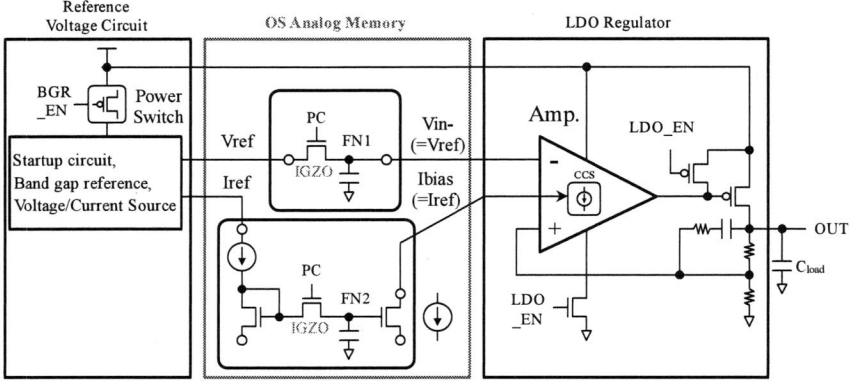

Figure 13. A circuit configurations of the integrated voltage regulator (IVR).

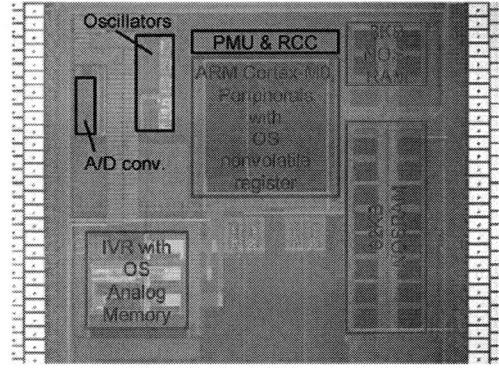

Figure 14. A chip micrograph of the 32-bit Noff microcontroller.

5 Application to AI accelerator

In the case where an AI accelerator performs operation, it usually requires access to large-scale memory such as external DRAM as shown in Fig. 15(a). The energy required for data transfer between a chip and external memory is high. Thus, in the case where the AI accelerator requires frequent memory access, its power consumption increases and its power efficiency decreases.

Our technical approach is to monolithically stack OS memory directly on a multiply-accumulate (MAC) array and directly connect the MAC array and the OS memory, as shown in Fig. 15(b). With this approach, the energy required for memory access is significantly reduced (to approximately 1/100) compared to the configuration of Fig. 15(a). Our approach to stack OS memory is particularly effective in an application that requires frequent memory access, such as an AI accelerator, and is an essential technique to achieve high power efficiency.

In addition, a network with high inference accuracy tends to have many layers and many weight parameters and thus needs large-scale memory. As described in Section 3, CAAC-IGZO FETs can be stacked, which increases the integration degree of memory in the 3D direction within a limited chip area. As a result, LSI using CAAC-IGZO is suitable for the increase in capacity of on-chip memory. In other words, the AI accelerator is one of the optimal applications that utilize the features of the CAAC-IGZO FET.

Examples of typical neural networks include a fully-connected neural network (FC) and a convolutional neural network (CNN). In the FC, the number of weight parameters is large, and the amount of MAC operations is equivalent to memory usage. In contrast, in the CNN, the same weight parameter is repeatedly used in operations while different data are input for the operation; thus, the CNN needs a smaller number of weight parameters and performs much larger amount of MAC operations than the FC. CAAC-IGZO technologies applicable to AI applications of FC and CNN will be described below.

(a)

(b)

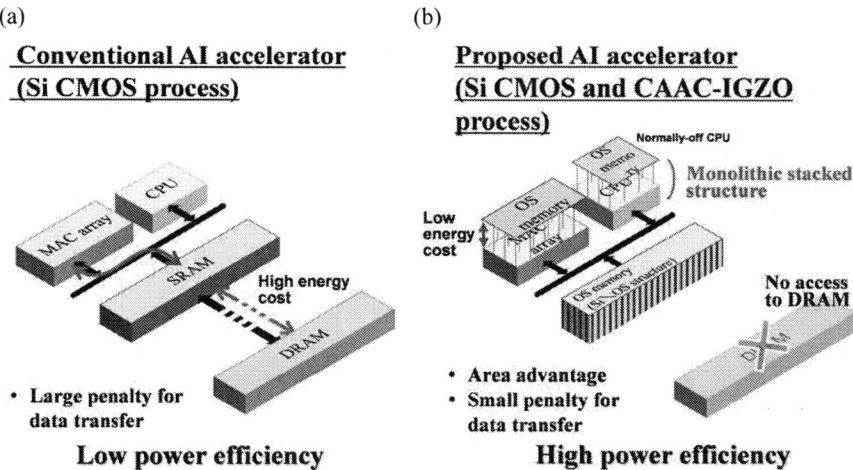

Figure 15. AI accelerator configuration examples: (a) a possible configuration example based on a Si CMOS process and (b) a configuration where Si CMOS and CAAC-IGZO are monolithically stacked.

5.1. Noff Processor with AI Accelerator for Fully-Connected Binary Neural Network

In a general fully-connected neural network, the number of parameters is large, and the amount of MAC operations is equivalent to memory usage. Thus, in the case where inference processing is performed in a circuit, matrix multiply architecture is adopted. When a Si substrate is only used, memory and an operation unit are fabricated in the same plane; therefore, the number of parallel processings is limited by the physical distance between the memory and the operation unit in the matrix multiply architecture.

Figure 16 shows circuit diagrams of an OS memory cell that constitutes an AI accelerator and a Si processing unit. The memory cell is composed of three FETs and one capacitor that are all formed using CAAC-IGZO layers. The memory cell retains charge at FN and thus loses no data even during a power-off period. Therefore, the memory cell is used as memory that retains learning data for inference through Noff driving by learned data storage. The processing unit is fabricated using Si CMOS and includes a multiplier, an adder, an accumulator, a batch normalization selector, and other components.

Figure 17 is a schematic diagram showing the layout of the OS memory and the processing unit. The OS memory array divided into four sections is provided for one subarray, and all the bit lines in the memory array are connected to the processing unit provided in a lower Si layer. In the entire AI accelerator, every 2k lines is connected to the processing unit. Data read from the bit lines in parallel are subjected to operations at the same time, as described above. As can be seen from Fig. 17, the processing unit is provided in the Si layer that is the lower layer of the memory cell fabricated using only CAAC-IGZO FET, which solves the issue of the physical distance. Furthermore, the lower layer of the memory cell also includes a column driver; therefore, the chip area is reduced by approximately 25%. Moreover, the bit line lengths can be finely divided for the memory subarrays like the DOSRAM, which contributes to reductions of memory read energy by approximately 70% in the case where the bit line lengths are divided into eight.

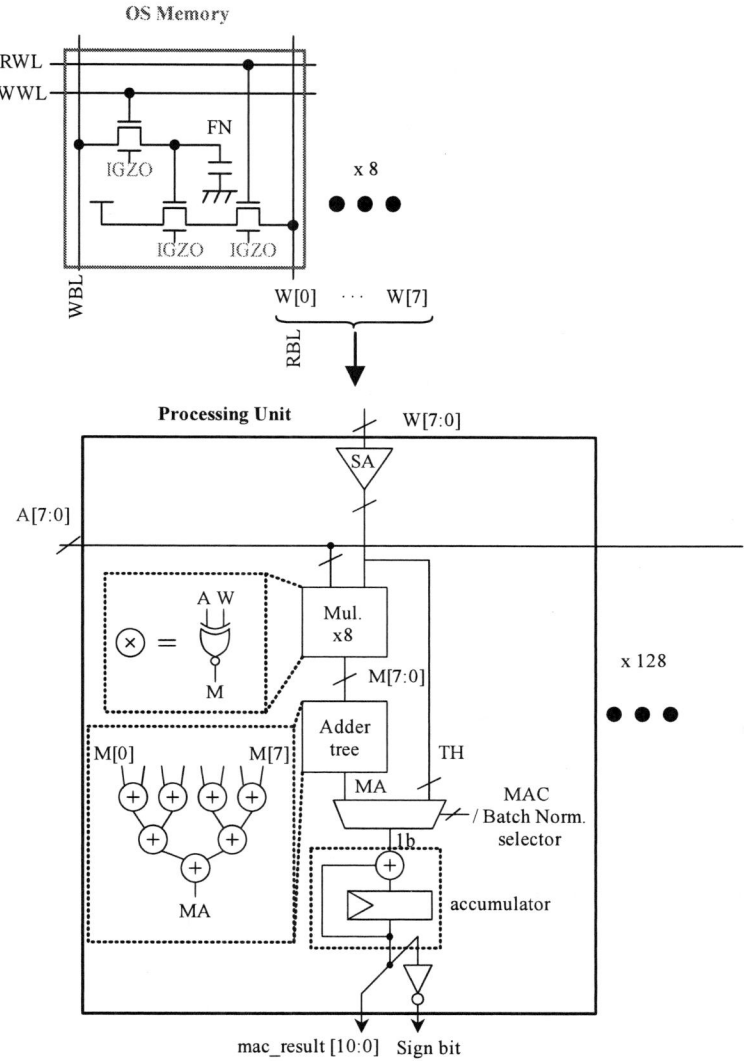

Figure 16. Diagrams of the OS memory cell that constitutes the AI accelerator and the Si operation circuit.

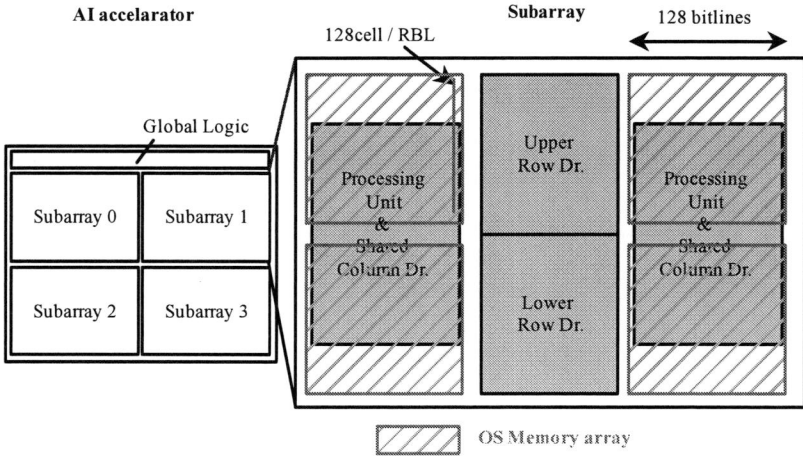

Figure 17. A top schematic diagram showing stack layout of the OS memory that constitutes the AI accelerator and the processing unit.

5.2. CAAC-IGZO AI Accelerator For Convolutional Neural Network With 8-bit Image Recognition

A neural network with image recognition includes multiple CNN layers. For accurate image recognition, the neural network requires 8-bit or more operation accuracy (39). Image recognition especially for automotive applications needs high-speed operation because images should be judged instantaneously. In addition, it is sometimes difficult to perform chip cooling. Thus, reducing power consumption is necessary in order to inhibit the increase in chip temperature. One of the approaches for achieving these is to provide an AI accelerator with power efficiency of, for example, higher than 100 TOPS/W. We have fabricated an 8-bit CNN CAAC-IGZO AI accelerator to achieve high performance.

A circuit block of processing elements (PE) of the CNN AI accelerator is shown in Fig. 18. In the circuit, a large number of OS memory cells are provided in an CAAC-IGZO layer to store weight parameters. The weight parameters can be read through local bit lines (LBL) into a read circuit in massively parallel. A Si layer mainly performs MAC operation using weight parameters read from the OS memory cells and post-processings of the MAC operation, such as scaling, activation, and pooling.

Since the CAAC-IGZO FET exhibits lower mobility and on-state current than Si CMOS, operation of reading the weight parameters from the OS memory cells might be a bottleneck in AI operation. The actual read speed of a NOSRAM cell composed of only CAAC-IGZO FETs is 45 ns (approximately 22 MHz) (29). However, the CNN has architecture where the same weight parameter is repeatedly used; thus, the interval between the operations of reading the weight parameters from the OS memory cell is long. In addition, the order of the operations of reading the weight parameters is determined in the AI operation, and thus the memory sequentially reads the weight parameters. From the above two reasons, it is possible to use a configuration as shown in Fig. 19 in which a large number of data are read from the OS memory cell into the LBLs in massively parallel and data required for the read circuit is selected at high speed and output to a global bit line (GBL). With this driving, the OS memory cell is not a bottleneck of the entire chip operation, and an operation unit fabricated using Si CMOS operates at 200 MHz or higher.

(a)   (b)

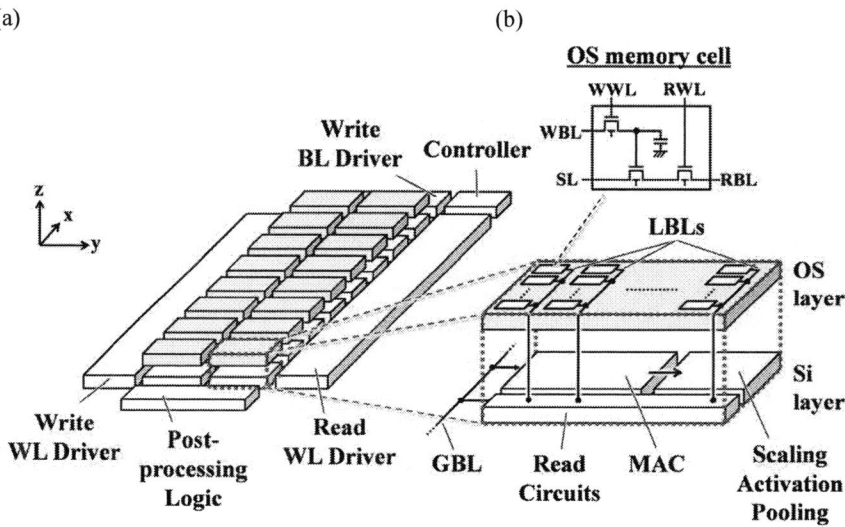

Figure 18. (a) A PE 3D schematic diagram and (b) a schematic diagram of a stack structure of MAC and OS memory.

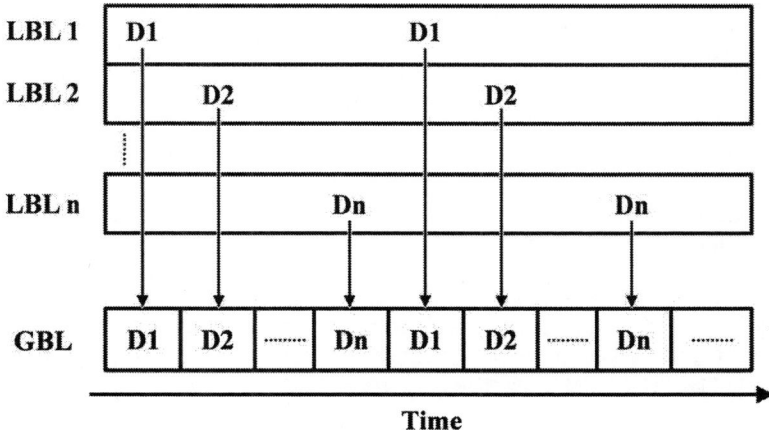

Figure 19. Timing of reading weight parameters from LBLs into GBL.

Table I shows design specifications of the CNN AI accelerator designed with a 60-nm CAAC-IGZO/55-nm Si CMOS hybrid process. In this configuration, the operation unit and the memory are stacked and their physical distance is short; thus, memory read energy is very low. With designed values, a power efficiency of 7.6 TOPS/W is estimated in each PE. Even when further scaling is conducted, the use of this configuration solves the memory access bottleneck and provides the effect of power consumption reduction due to scaling (40). On the assumption that Si CMOS and CAAC-IGZO are scaled down to 14 nm or less in accordance with Reference (40), a power efficiency of higher than 100 TOPS/W (higher than 8-bit power efficiency) would be achieved, as shown in Table II. Moreover, on the assumption of a 1-bit binary network and scaling up to 7 nm, a power efficiency of higher than 1 POPS/W (higher than 1-bit power efficiency) would be achieved.

TABLE I. Design specifications

	Designed value
Process node	CAAC-IGZO: 60 nm Si CMOS: 55 nm
Function	CNN
Core area	6.25 mm^2 (2.5mm × 2.5mm)
OS memory capacity (for weight parameter storage)	58.5 kB
Operating frequency	200 MHz
Data Type	INT8
Peak performance (per PE)	13 GOPS at 200 MHz
Power efficiency (per PE)	7.6 TOPS/W at 1.2 V

TABLE II. AI accelerator power efficiency

	Designed value	When scaled down (assumption)		
Process node	55 nm	28 nm	14 nm	7 nm
Power supply voltage	1.2 V	0.6 V	0.6 V	0.6 V
Power efficiency (8-bit, in each PE)	7.6 TOPS/W	60 TOPS/W	120 TOPS/W	0.24 POPS/W
Power efficiency (1-bit, per PE)	53 TOPS/W	0.42 POPS/W	0.83 POPS/W	1.7 POPS/W

5.3. Expectation of Analog CIM using CAAC-IGZO FETs

Computing-in-memory (CIM) where MAC operation is performed in a memory circuit would achieve high power efficiency (41). Like emerging nonvolatile memories, memory using CAAC-IGZO FETs is one of the candidates for devices that achieve CIM (42), and we have reported a fabricated analog multiplication operator that exhibits good multiplication, retention, and temperature characteristics and small device variations (43). In order to achieve high power efficiency, analog circuits using CAAC-IGZO FETs as well as digital circuits and memory using CAAC-IGZO FETs will be key devices.

6 Conclusion

We have proposed crystalline oxide ceramics called CAAC-IGZO as a next-generation material in CMOS LSI, which is currently a mainstream, in order to reduce a huge amount of power consumption, which might be a big issue in the AI era. A CAAC-IGZO FET exhibits extremely low I_{off} in the order of yA/μm (10^{-24} A/μm) and is expected to be applied to memory and a Noff-CPU owing to its unique characteristics.

In an application of a CAAC-IGZO FET that can be fabricated in Si CMOS back end of line processes to an AI accelerator, no access to external memory will eliminate a memory access bottleneck and offer possibilities of 100 TOPS/W with 8-bit accuracy and 1 POPS/W with 1-bit accuracy, as shown in Fig. 20 (44). The combination of these CAAC-IGZO technologies would reduce power consumption at the system level and solve the issue of power consumption in the AI era.

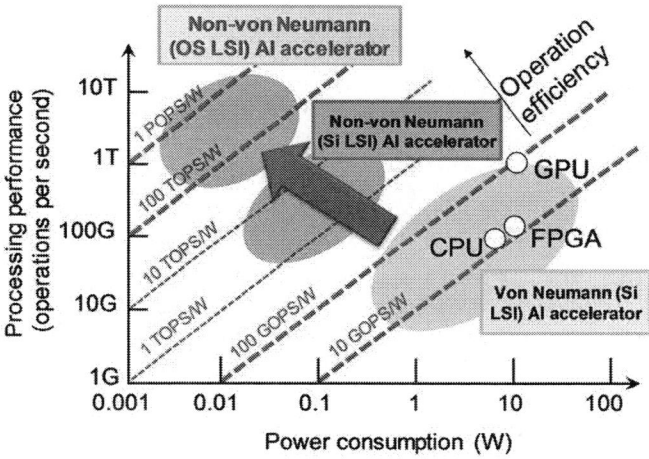

Figure 20. Power consumption and power efficiency of AI accelerators.

References

1. Ministry of the Environment, Government of Japan, "Global Warming and Infectious Diseases: What we know now?," 2007 [published in Japanese]. https://www.env.go.jp/earth/ondanka/pamph_infection/full.pdf
2. Center for Low Carbon Society Strategy, Japan Science and Technology Agency, "Impact of Progress of Information Society on Energy Consumption (Vol.1)," 2019 [published in Japanese]. https://www.jst.go.jp/lcs/pdf/fy2018-pp-15.pdf
3. S. Yamazaki, "Challenge of crystalline IGZO ceramics to silicon LSI", speech at Science Seminar 2018 held by Japan-Sweden Foundation at Embassy of Sweden, Alfred Nobel Auditorium on November 26, 2018.

4. S. Yamazaki, "Physics and Technology of Crystalline Oxide Semiconductor CAAC-IGZO: Fundamentals", John Wiley & Sons, 2017.
5. Y. Waseda et al., *Mater. Trans.,* **59**(11), 1691 (2018).
6. S. Yamazaki et al., *Jpn. J. Appl. Phys.*, **55**, 115504 (2016).
7. S. Yamazaki et al., *J. Soc. Inf. Disp.*, **22** (1), 55 (2014).
8. S. Ito et al., *Proc. AM-FPD '13*, 151 (2013).
9. M Takahashi et al., Semiconductor Energy Laboratory Co., Ltd., Semiconductor device, Japanese Patent JP 5894694. Mar 30, 2016.
10. S Yamazaki et al., Semiconductor Energy Laboratory Co., Ltd., Oxide semiconductor, United States Patent US 9,153,652. Oct 6, 2015.
11. S. Yamazaki et al., *ECS Trans.*, **67**(1), 29 (2015).
12. D. Matsubayashi et al., *IEDM Tech. Dig.*, 141 (2015).
13. H. Kunitake et al., *IEDM Tech. Dig.*, 312 (2018).
14. H. Kunitake et al., *IEEE J-EDS*, **7**, 495 (2019).
15. A. Suzuki et al., *PRiME 2020*, to be published.
16. K. Kato et al., *Jpn. J. Appl. Phys.*, **51**, 021201 (2012).
17. S. Yamazaki et al., *Int. J. Ceramic Eng. Sci.*, **1**(1), 6 (2019).
18. E. Morifuji et al., *Symp. VLSI Tech. Dig.*, 163 (1999).
19. H. Sunamura et al., *Symp. VLSI Tech. Dig.*, T250 (2013).
20. M. Oota et al., *IEDM Tech. Dig.*, 50 (2019).
21. J. Wu et al., *Symp. VLSI Tech. Dig.*, THL.4 (2020).
22. H. Fujiwara et al., *Symp. VLSI Tech. Dig.*, TH2.2 (2020).
23. S. Yamazaki, "Physics and Technology of Crystalline Oxide Semiconductor CAAC-IGZO: Application to Display", John Wiley & Sons, 2017.
24. S. Amano, et al., *SID2010 Digest*, PP. 626-629 (2010).
25. S. Katsui et al., *J. Soc. Inf. Disp.*, **27**(8), 497 (2019).
26. S. Yamazaki and M. Fujita, "Physics and Technology of Crystalline Oxide Semiconductor CAAC-IGZO: Application to LSI", John Wiley & Sons, 2017.
27. T. Onuki et al., *IEEE J. Solid-State Circuits*, **52**(4) 925 (2017).
28. G. S. Kar and A. Furnemont, *Solid State Technology*, **61**(7), 14 (2018).
29. S. Maeda et al., *Int. Solid-State Circuits Conf. Dig. Tech. Pap.*, 484 (2018).
30. T. Ishizu et al., *Symp. VLSI Tech. Dig.*, C162 (2017).
31. K. Ando et al, Nonvolatile magnetic memory, FED Journal, vol. 12, no. 4, 2001, pp. 89-95 (in Japanese)
32. N. Sakimura et al., *IEEE J. Solid-State Circuits*, **44**(8) 2244 (2009).
33. K. Nomura et al., *J. Appl. Phys.*, **111**, 07E330 (2012).
34. T. Ishizu et al., *SSC-L*, **2**(12), 293 (2019).
35. T. Ohmaru et al., *Ext. Abstr. Solid State Dev. Mater.*, 1144 (2012).
36. H. Kobayashi et al., *IEEE CooL Chips XVI,* Session VI (2013).
37. A. Isobe et al., *Symp. VLSI Tech. Dig.*, 49 (2014).
38. H. Tamura et al., *IEEE COOL Chips XVII,* Session XII (2014).
39. B. Moons et al., *IEEE WACV 2016*, pp. 1-8 (2016).
40. A. Stillmaker and B. Baas, *INTEGRATION the VLSI journal*, **58**, 74 (2017).
41. H. Valavi et al., *Symp. VLSI Tech. Dig.*, 141 (2018).
42. S. Cosemans, *Int. Solid-State Circuits Conf. Forum F2*, (2019).
43. Y. Kurokawa et al., *Jpn. J. Appl. Phys.*, **59**, SGGB03 (2020).
44. M. Hayashikoshi, *Int. Solid-State Circuits Conf. Forum F2*, (2019).

ECS Transactions, 98 (7) 205-217 (2020)
10.1149/09807.0205ecst ©The Electrochemical Society

Sub-40mV Sigma-VTH IGZO nFETs in 300mm Fab

Jerome Mitard, Luka Kljucar, Nouredine Rassoul, Harold Dekkers, Michiel van Setten,
Adrian Vaisman Chasin, Geoffrey Pourtois, Attilio Belmonte, Gabriele Luca Donadio,
Ludovic Goux, Ming Mao, Harinarayanan Puliyalil, Lieve Teugels, Diana Tsvetanova,
Manoj Nag, Soeren Steudel, Jose Ignacio del Agua Borniquel[2a], Jothilingam
Ramalingam[2], Romain Delhougne, Chris J.Wilson, Zsolt Tokei, Gouri Sankar Kar

imec, Kapeldreef 75, B – 3001 Leuven, Belgium
[2] Applied Materials, MPD, Santa Clara, USA
[2a] also imec resident, Kapeldreef 75, B – 3001 Leuven, Belgium

> Back and double gate IGZO nFETs have been demonstrated down
> to 120nm and 70nm respectively leveraging 300mm fab processing.
> While the passivation of oxygen vacancies in IGZO is challenging
> with an integration of front side gate, a scaled back gated flow has
> been optimized by multiplying design of experiments around
> contacts and material engineering. We then successfully
> demonstrated sub-40mV $\sigma(V_{TH\_ON})$ in scaled IGZO nFETs.
> Regarding the performance and the V_{TH_ON} control, a new IGZO
> phase is also reported. A model of dopants location is proposed to
> better explain the experimental results reported in literature.

Introduction

In today's highly information-oriented society, groundbreaking hardware innovations
targeting Augmented Intelligence are necessary to compute on even-more limited power
budget. This requires higher density memory as well as smart interconnect solutions that
can actively reconfigure leveraging the 3D dimension [1]. InGaZnO4 (IGZO) as a thin film
channel material for FETs can become a key element for the aforementioned applications
since its premise relies on 1) zepto ampere off-state leakage capability namely few
electrons flowing per FET element in one year [2], 2) relatively good electron mobility
especially when compared to that of doped-amorphous Si [3] and 3) low thermal budget
processing to possibly enable a sequential integration with conventional Si-based
transistors. In this work, IGZO device integration is reported leveraging our 300mm-fab
facilities. Our objective is mainly to gain insights into the process and material elements
which drive the control of the performance parameters of IGZO nFETs.

Fabrication Of Scaled 300mm-IGZO NFET

Process details are given in Figure 1. IGZO nFET consists of three main modules: active
area, source/drain contacts and front side gate in case of double gated nFETs. First, a highly
doped Si 300mm wafer with an option of blanket metal deposition serves as a common
back-gate for the transistors. A thick (SiCN/) ALD Al2O3 is used as bottom gate dielectric.
The next step utilizes 300mm AMAT Endura® Impulse PVD IGZO. Active patterning is
applied on the entire stack down to the Si substrate. The gate is formed by patterning

TiN/W stack down to Al_2O_3 gate dielectric level. This minimizes the damages induced by gate patterning. The last module is the Source and Drain contact formation through a PECVD SiO_2 field oxide. Ti-based/W metallization is applied similarly to [4].

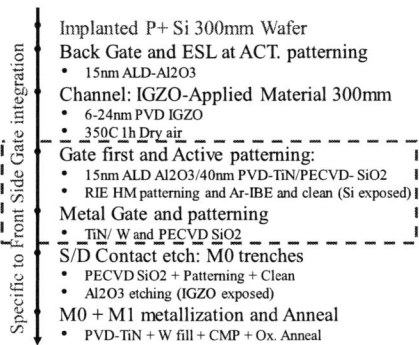

Figure 1. Detailed flow description of back and double IGZO nFETs.

It is worthwhile mentioning that devices have been successfully demonstrated up to M1 level enabling the study of small device dimensions. Figure 2 Left shows fully functional 70nm-Lg double gate and 120nm-Lg back gate IGZO nFETs. Corresponding xTEM images are given in Figure 2 -Right panel as reference.

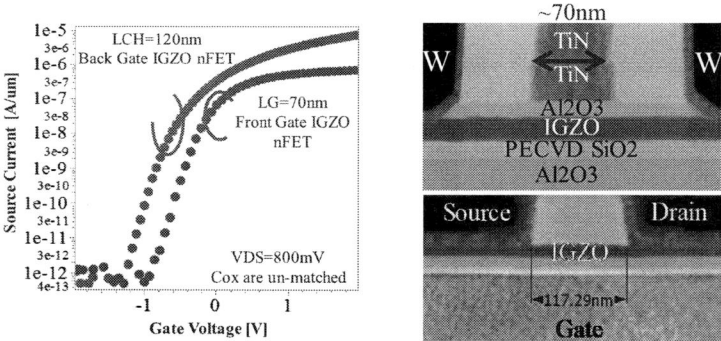

Figure 2. Left. IVs of 70nm-front side and 120nm back-gate IGZO nFETs. Right-top: xTEM of 70nm-front side IGZO nFETs. Bottom-Right: xTEM of 120nm-Back gate IGZO.

Discussion About Variability In IGZO NFET

N-type doping depends on the IGZO crystallographic structure

There are two main electron doping mechanisms known in amorphous-IGZO: a sub stoichiometric amount of oxygen [oxygen vacancies (Vo)] and the incorporation of hydrogen. Both give very shallow donor levels in the bandgap and then act as n-type dopants in IGZO. The weakest bonded oxygen atoms like those that are under-coordinated or that are involved in multiple Zn-O bonds [5] can be released in presence of hydrogen to form OH ions [6]. Fig. 3 shows the impact of a mild 10% H anneal at >200°C.

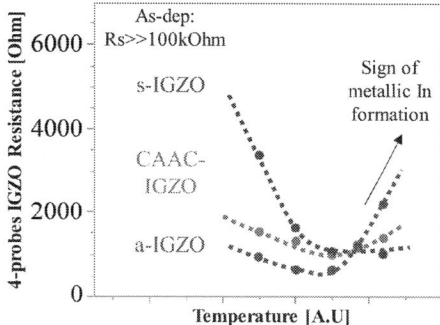

Figure 3. Resistivity measurements after 1h FGA (10% H_2). 50nm IGZO different phases and temperatures.

A clear drop of the IGZO resistivity with respect to the control samples (as-deposited) is in line with a doping increase. More interestingly, we found that the initial response to hydrogen anneal can be modulated by the crystallographic structure of IGZO.

The challenge of passivating oxygen vacancies w. top gate

To control the final doping of the IGZO channel, it is well reported in literature [7] that a post processing oxygen anneal can be applied to passivate the oxygen vacancies which are formed during the device fabrication. This technique is also used in our 300mm flow as described in Fig. 1. As it might be expected, in front gate IGZO nFETs, the passivation efficiency is limited by the presence of the top gate stack. Therefore, it seems important to limit the formation of the oxygen vacancies during the deposition of the gate dielectrics. Three oxides were studied: SiO_2, Al_2O_3 and HfO_2. The experiment consists of measuring the IGZO resistance before and after oxide deposition. Figure 4 confirms that a low resistivity IGZO is measured after oxide deposition. After oxygen annealing, only PECVD-SiO_2 shows a large recovery while for Al_2O_3 and HfO_2, it remains low.

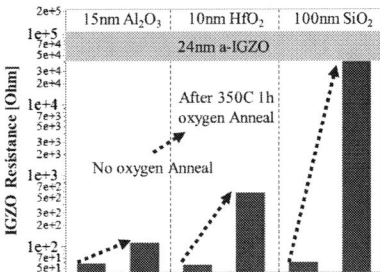

Figure 4. 4p-Resistance measurement of CTLM devices (10um-spacing). Effect of the oxygen Anneal in presence of IGZO capping.

This result clearly challenges the implementation of conventional water-based high-K materials in gate first IGZO integration. However, a high I_{ON}/I_{OFF} ratio can still be obtained with highly doped (i.e. un-passivated Vo) IGZO nFETs by increasing the gate coupling. Fig. 5 shows an experimental demonstration of a 6nm-thick doped-IGZO nFETs down to 120nm-Lg. Since ultra-thin IGZO integration has its own challenge in term of variability and process controllability, we primarily focus on back gated IGZO nFET from now on.

Figure 5. Effect of IGZO thickness on electrostatic control. Full carrier depletion can be obtained.

Contact engineering for IGZO nFET with Ti metal barrier

Contrary to the use of undoped IGZO in the channel for I_{OFF} control, contacts can rely on maximizing the oxygen vacancies to increase the dopant concentration in the S/D regions. When not done locally, it could be an extra source of variability. This doping is made through oxygen scavenging from IGZO by a thin Ti layer between the liner (TiN) and the IGZO. Low specific contact resistivity down to 1×10^7 Ohm.cm² is demonstrated when the Ti layer is thinner than 5nm (Fig. 6). With thicker barrier, the formation of TiO_2 and specific alloys is taking place at the IGZO/Contact interface. This is both confirmed by ab initio simulations (Fig. 7) and by advanced physical characterization techniques (see Fig. 8).

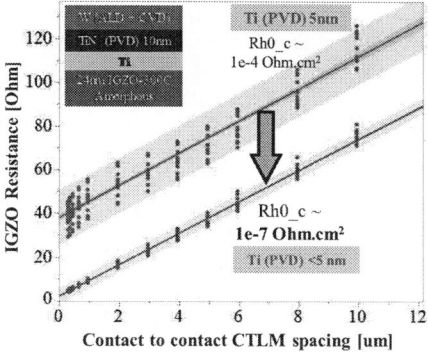

Figure 6. IGZO contact engineering with Ti thickness optimization.

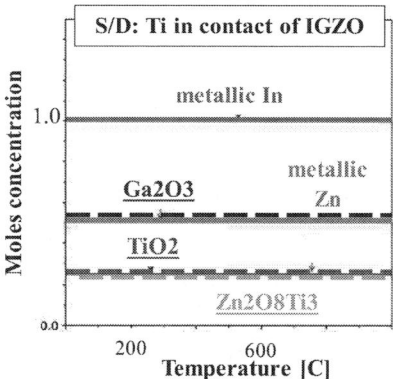

Figure 7. Ab-Initio (DFT@PBEsol) modeling of the IGZO/Ti interface.

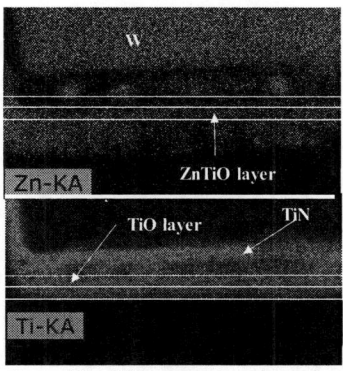

Figure 8. EDS characterization of the S/D contact region when 10nm-thick Ti is in contact with IGZO-PVD.

Amorphous IGZO, C-Axis Aligned IGZO and new s-IGZO

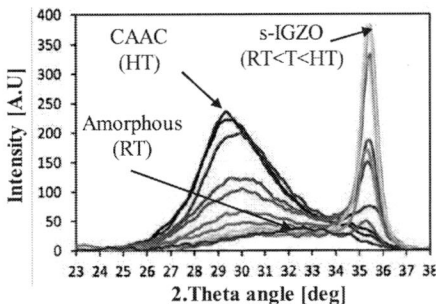

Figure 9. XRD characterization of blanket IGZO when chuck temperature is varied from Room Temperature to High Temperature.

X-ray diffraction (XRD) techniques are used extensively since this is a nondestructive in-fab technique that provides detailed information about the crystallographic structure of the IGZO material. Figure 9 shows a typical spectrum where clear peaks/humps are seen and attributed to: 1. amorphous IGZO, 2. CAAC-IGZO and 3. to the best of our knowledge, previously not reported new phase called here s-IGZO (spinel phase). The s-phase is only formed under certain conditions of power, temperature, and oxygen flow during material deposition. In parallel, we verified that the transition between the different phases is not due to any compositional changes within the IGZO material.

Thick IGZO (>12nm) backgated nFETs with active layers submitted to final O_2 anneal are used to study the different phases of IGZO and their electrical impact on device parameters. In this configuration, the carrier transport preferentially occurs in the bottom half of the IGZO channel while the top half (SiO_2/IGZO interface) mostly drives the electrostatic control of BG transistors. In Fig.10, I_{ON}-V_{TH_ON} long channel trends are shown and highlight a clear difference between CAAC- and a-IGZO already.

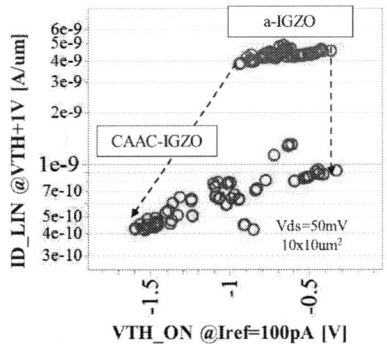

Figure 10. Degradation of ID in linear regime, taken at offset V_{TH_ON} and increased variability for 24nm-CAAC-IGZO.

The amorphous IGZO has much reduced spread in VTH-ON and higher ID,LIN (~mobility) than CAAC-IGZO. Reliability tests have also been carried out to compare the two phases and the results show the existence of two competing PBTI degradation mechanisms for CAAC-IGZO (Figure 11). Notice that it is important to report the degradation over time and different gate voltage stress because a cut line at a specific value could have shown no BTI degradation of CAAC-IGZO and then attributing an unfair benefit of this phase over the a-IGZO phase. Combining these findings to the result shown in Fig. 3 where the sheet resistance of IGZO under hydrogen exposure depends on the phases, we can reasonably conclude that CAAC owns different doping levels in comparison to a-IGZO.

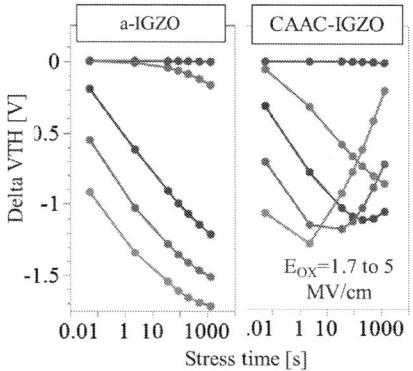

Figure 11. Positively-charged trapping only observed in CAAC-IGZO ("Si-like" PBTI).

To gain insight into longitudinal doping (from Source to Drain), a short channel study is performed. The impact of an increased oxygen partial pressure during the PVD deposition is first reported for both a-IGZO and CAAC-IGZO. Figure 12 - left side - demonstrates that a constant increase of the oxygen partial pressure during amorphous-IGZO deposition results in a continuous increase of I_{ON} and V_{TH_ON}. This is in line with a reduction of oxygen vacancies and therefore a lowering of effective n-type doping in the thin film [7]. On the contrary, CAAC-IGZO formed at temperature larger than 25°C and for a positive O_2% flow show reversed trends namely a reduced I_{ON} as well as V_{TH_ON} from low to high oxygen content (Fig. 12 -Right panel-).

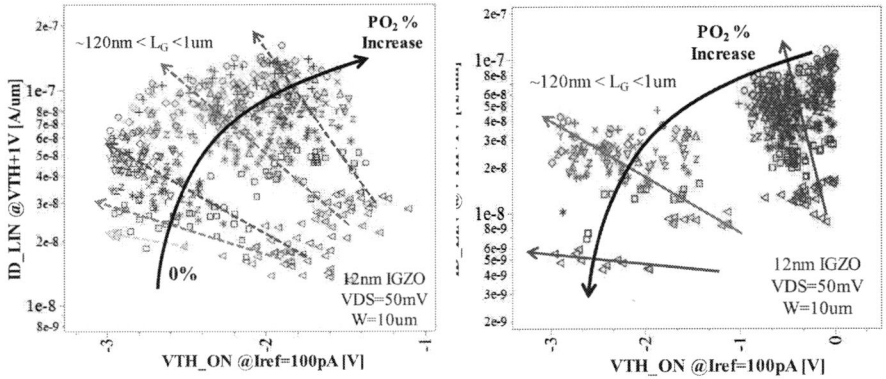

Figure 12. (Left) Increased flow of oxygen deposition improves VTH_ON and Idlin performance of a-IGZO. (Right) Increased flow of oxygen during deposition degrades VTH_ON and Idlin performance of CAAC-IGZO

In CAAC-IGZO, supplying oxygen in excess leads to a severe degradation of the performance of the transistors, especially when compared to Room Temperature (RT) amorphous IGZO. The temperature during IGZO deposition can also be tuned to obtain s-IGZO. Fig.13 shows that short channel effect control of s-IGZO nFETs have the best trade-off between I_{ON} (mobility) and V_{TH_ON} (doping) over the two other IGZO structures (CAAC and a-IGZO).

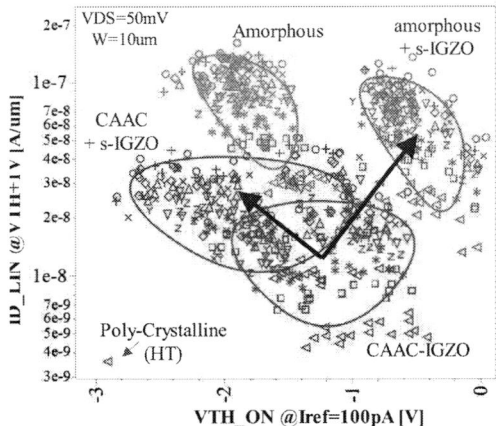

Figure 13. The combo of 12nm- aIGZO and the new s-IGZO gives the best trade-off VTH_ON short channel performance.

Increased mobility in s-IGZO has been studied by ab-initio simulations and we confirm that this specific crystalline structure leads to free carriers with lower effective mass and consequently higher mobility than with CAAC-IGZO (see Fig. 14).

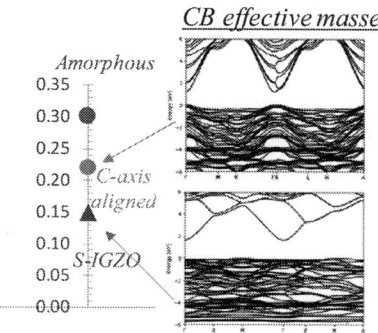

Figure 14. Ab-initio simulations confirming the s-phase of IGZO has reduced effective mass as compared to CAAC. The effective mass of the two crystalline phase is obtained from the calculated band-structure, the effective mass for a-IGZO is taken from literature [8].

Since the back gated architecture allows a profiling of dopant within IGZO, a sequential deposition of a-IGZO and semi-crystalline IGZO has been attempted to see if both the carrier transport and the short channel effects control can be improved. A 6nm amorphous layer followed by a 6nm low 0%-CAAC IGZO should bring the short channel IGZO nFETs to the high-performance side. However, depositing the CAAC IGZO layer on the amorphous phase IGZO causes severe device degradation for all different layer thickness ratios (see Fig.15). This result is explained by the thermal budget used to create the CAAC structure which is incompatible with RT a-IGZO. One-step s-IGZO thin-film deposition seems an attractive option for obtaining high performance short channel IGZO nFETs.

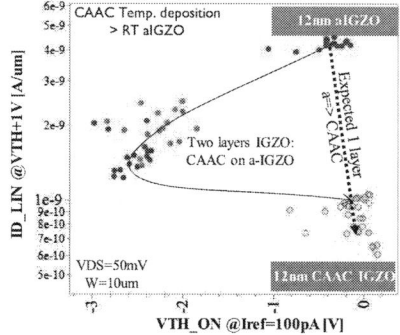

Figure 15. Dual IGZO deposition CAAC on top of a-IGZO. Higher thermal budget of the second layer degrades the bottom a-IGZO.

The lateral dopant profiling is done through the study of the effect of narrowing the channel width of IGZO. 20nm-thick low O% CAAC IGZO nFET is selected because we revealed moderate V_{TH_ON} (LCH) roll-off, meaning a reduced amount of Vo in the top half of the IGZO channel if compared to a-IGZO. Fig. 16 shows the effect of WCH scaling on V_{TH_ON} where a drastic improvement of electrostatic control is observed for the transistor width smaller than 200nm. This result highlights a strong non-uniformity of n-type dopant concentration along the width of the device.

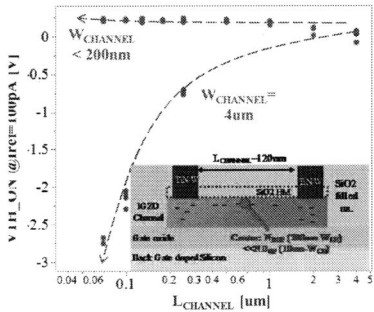

Figure 16. Narrowing the IGZO channel drastically increases the controllability of V_{TH_ON}.

A Scanning Spreading Resistance Microscopy (SSRM) analysis has been performed on test samples (Fig. 17) and a reduced resistivity is observed near the edge of the IGZO active area. Note that the effect cannot be seen along $L_{CHANNEL}$ due the high n-type doping created by the S/D metallization. An empirical model has been built to gather all the information from the previous experiment (Fig. 17).

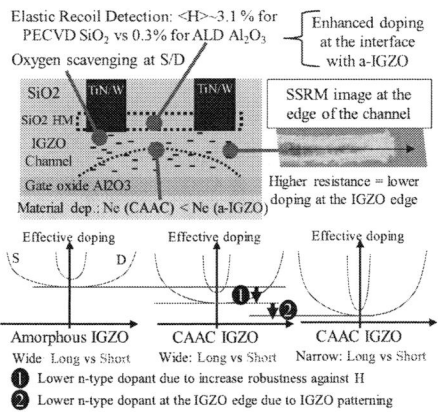

Figure 17. Dual IGZO deposition CAAC on top of a-IGZO. Higher thermal budget of the second layer degrades the bottom a-IGZO.

214

Demonstration Of σ(VTH-ON) Down To 20mV

Based on the previous learning about n-type dopant location in IGZO, we show that the performance of transistors keeps increasing when the channel length is reduced and scales with the channel width (Fig. 18). A measure of the V_{TH-ON} variation has been performed at the 300mm wafer scale from long to ~120nm L_{CH} and down to 200nm W_{CH} dimensions. Fig. 19 shows the Id-Vg curves of >100 Back Gated IGZO-nFETs with no failed devices detected. The standard variation of the V_{TH-ON} across L_{CH} and W_{CH} is often less than 40mV with a minimum of 20mV (Fig. 20). The NBTI reliability has been evaluated on these Vo-controlled back gated IGZO nFETs. Limited V_{TH-ON} degradation under NBTI stress has been detected up to 1000s stress time and at oxide field up to 5MV/cm (Fig. 21).

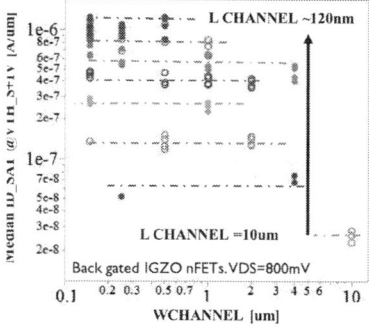

Figure 18. I_{D-SAT} at offset V_{TH-SAT} keeps increasing with $W_{CHANNEL}$ is scaled down.

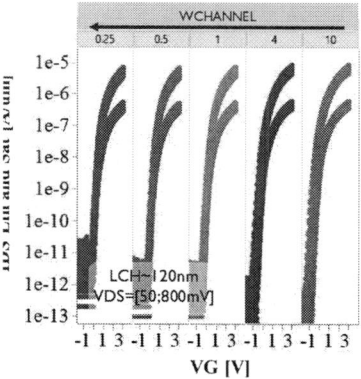

Figure 19. More than 100 back-gated IGZO nFETs functional across W_{CH} dimensions. No filtering process applied.

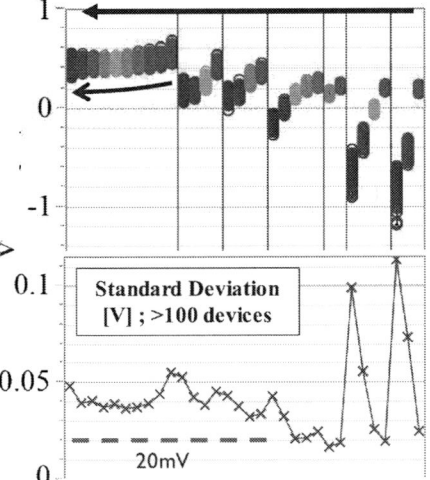

Figure 20. More than 100 back-gated IGZO nFETs functional across W_{CH} dimensions. No filtering process applied.

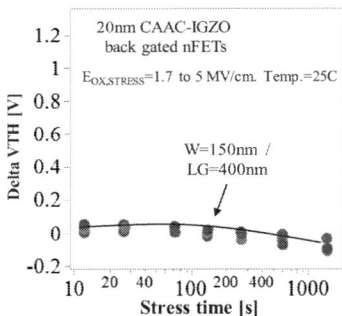

Figure 21. Stress and Sense NBTI degradation on narrow IGZO nFETs at V_{TH_ON} overdrive up to 11V.

Conclusions

We have demonstrated scaled IGZO nFETs with excellent V_{TH_ON} control using an industry compatible 300mm process flow. This has been achieved thanks to a careful mapping of n-type doping in the three dimensions of the IGZO channel. While semi-crystalline IGZO seems to be more robust against hydrogen than that of amorphous-IGZO, a new IGZO phase is found to help boost I_{ON} at short channel. The demonstration of back-gate IGZO nFETs with low variability and relatively high drive will benefit to the top-gate architecture development. This will provide new opportunities for the IGZO-based devices like an "on-the-fly" V_{TH} setting for the performance control of advanced applications.

Acknowledgements

The imec's Logic and Memory partners involved in the NanoIC and Emerging Memory programs, EU (project funded under the grant agreements: 687299 (NeuRam3-Horizon 2020) and 826655 (Tempo-Electronic Components and Systems for European Leadership Joint Undertaking), the pilot line and amsimec (test lab) are acknowledged for their support. A special thank you go to the numerous colleagues from the Unit Process and MCA teams as well as Y Cao & D. L.Diehl from Applied Materials Inc. Si Syst., MPD, Sunnyvale and A.Cockburn, AMAT-Belg. This work is a broad team effort.

References

1. Paul Heremans, *Imec Technology Forum*, https://2019.futuresummits.com/itf2019/belgium, (2019).
2. Shunpei Yamazaki, "Challenge of crystalline IGZO ceramics to silicon LSI - Its application to AI and displays" in "Semiconductor Technology for Ultra Large Scale Integrated Circuits and Thin Film Transistors VII", *ECI Symposium Series*, (2019)
3. F. Mo, "Experimental Demonstration of Ferroelectric HfO2 FET with Ultrathinbody IGZO for High-Density and Low-Power Memory Application," *2019 Symposium on VLSI Technology*, Kyoto, Japan, 2019, pp. T42-T43, doi: 10.23919/VLSIT.2019.8776553.
4. L. Kljucar, "IGZO integration scheme for enabling IGZO nFETs", Thin Film Electronics: Oxide, Non-single Crystalline and Novel Process, *International Conference on Solid State Devices and Materials*, Nagoya, pp. 303 (2019).
5. N. Saito, "High-Mobility and H2-Anneal Tolerant InGaSiO/InGaZnO/InGaSiO Double Hetero Channel Thin Film Transistor for Si-LSI Compatible Process." *IEEE Journal of the Electron Devices Society*, 6, 500-505 (2018).
6. SW. Kong, "TCAD Simulation of Hydrogen Diffusion Induced Bias Temperature Instability in a‐IGZO Thin‐Film Transistors", *SID Technical digest*, (2017).
7. Kimizuka and Yamazaki, "Physics and technology of crystalline oxide semiconductor CAAC-IGZO", *Wiley* (2017)
8. G. Pourtois, *Phys. Rev. B* **75**, 035212 (2007)

Hafnia Ferroelectric Device for Semiconductor, Display and Sensor Applications

Sanghun Jeon*

School of Electrical Engineering, Korea Advanced Institute of Science & Technology, Daejeon, 34141, Korea, *jeonsh@kaist.ac.kr

For the ferroelectric material with the conventional perovskite crystal structure and the chemical formula of ABO_3, the charge neutral level is located close to the conduction band edge, and when forming a device, it causes a high leakage current due to low Schottky barrier height. Therefore, a relatively thick ferroelectric thin film was required, which was not suitable for the existing CMOS technology that is evolving through miniaturization. Meanwhile, ferroelectric properties were discovered in hafnia in 2011, and the ferroelectric properties of hafnia is observed in an orthorhombic crystal structure. Due to its simple structure and excellent CMOS process compatibility, it has received much attention for ferroelectric hafnia thin films and various application technologies using them. In this report, we will look at various application technologies using hafnia ferroelectrics, and we will examine the characteristics of hafnia ferroelectrics suitable for each device.

Introduction

Ferroelectrics are one of the most crucial inorganic materials due to their fantastic functionalities and their remarkable properties in response to electrical and mechanical stimuli [1-4]. As a representative ferroelectric material, the perovskite-type ferroelectrics materials have received intense attention. However, the electrical properties of perovskite materials become deteriorating with the scaling of area and thickness, which doesn't allow to make use it for a scaled device [1-2]. In particular, the process incompatibility between Si technology and the perovskite material has not been overcome and their applications are limited. Thus, the exploration for ferroelectric materials having outstanding process compatibility with Si technology has been highly required [3-4].

In 2011, hafnia-based ferroelectric material was reported and the most surprising feature of this material is its simple chemical composition and thermodynamic stability [5-6], which has outstanding process compatibility with Si-based semiconductor technology [1]. Hafnia-based materials are actually being utilized as alternative gate dielectric layers in MOSFET because of their high-κ value. The ferroelectric field effect transistor (FeFET) is being considered as the DRAM cell in next generation [7-8]. Multi-level FeFETs are reported not only for NAND flash but also for processing-in-memory architecture [9]. In addition, recent intensive studies have revealed new ferroelectric applications including sensor, logic and energy devices. In this presentation, we will review various device applications of hafnia ferroelectric thin films.

Results and discussion

NAND Flash with hafnia ferroelectric and thin film transistor

Recently, the flash-based memory cells have led revolution in computing memory and make deep in-road into the machine learning [9]. The flash memory has been continually scaled to small dimensions in order to meet the demand for enormous memory capacities.

Furthermore, the introduction of Multi-Level Cells (MLC) and 3-D vertical structure have improved density ahead of Moore's law. However, while memory density continues to improve, reliability characteristics such as endurance and retention are degraded. As a result, building larger-capacity and reliable flash memory will be getting harder.

In order to overcome these problems, we propose ferroelectric material based nonvolatile flash memory device as seen in Fig. 1, exploiting ferroelectric gate stack and thin film transistor. Hafnia based ferroelectric material has been promising because of the CMOS compatibility, impressive endurance ($>10^{12}$), and fast switching speed property ($<$50ns). The highly endurable ferroelectric flash can be an alternative to the conventional NAND flash with poor endurance performances. Moreover, ultra-fast memory device can be realized with ultra-short pulse switching property with less than nano seconds. We will show you how FeFET works and how this exciting new-comer' applies to next generation memory landscapes.

Figure 1. Schematic of vertical NAND Flash with hafnia ferroelectric materials

High-κ morphotropic phase boundary

Memory technology has evolved significantly over the past decades as it stores and processes increasing data. DRAM has one of the main memories that currently make up the computer memory hierarchy for some time now and will likely keep this position in the future as there are currently no emerging memories currently available for replacement. Memory devices other than DRAM have not been identified to simultaneously have high speed ($\sim$20 ns), high endurance ($>10^{15}$ cycles), high density ($>$several tens of gigabit), and a low cost. In order to continue the scaling of DRAM cell, it requires high dielectric constant dielectric formed in 3D structure as seen in Fig. 2 (a).

A ferroelectric material is one that generates a dipole switching in response to a electrical field. The most useful ferroelectric materials display a transition region in their composition phase diagrams, known as a morphotropic phase boundary [10], where the crystal structure changes abruptly and the ferroelectric properties are maximal. The utilization of the anisotropic phase boundary (MPB) between the tetragonal phase and the ferroelectric orthorhombic phase (see Fig. 2 (b)) newly discovered in hafnia is proposed for super highest dielectric constant and lowest equivalent oxide thickness capacitors. In our work, we experimentally demonstrate the MPB hafnia thin film with ~4.8 Å-EOT, extremely high dielectric constant of ~49 and sufficiently low leakage current ($<10^{-7}$ A/cm^2) for next generation DRAM device applications.

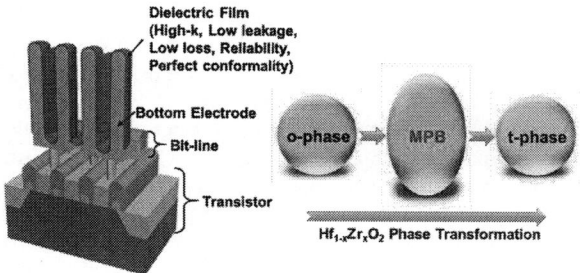

Figure 2. (a) Schematic of 1T-1C DRAM and (b) hafnia phase transformation

Ferroelectric transistor as for memory

Ferroelectrics are a kind of material composed of crystals that exhibit spontaneous electrical polarization. Ferroelectric device can be in multi-states states and manipulated with an external electric field [11]. When such an electric field is applied, the electric dipole formed from the crystal structure of the ferroelectric material tends to align with the electric field direction. After the field is removed, it remains polarized, giving the material non-volatile properties. The ferroelectric material has a nonlinear relationship between the applied electric field and the polarization charge, and provides a form of hysteresis loop for the ferroelectric polarization voltage (P-V) characteristic. Ferroelectric materials can be used as dielectrics in capacitors to be applied to memory applications such as DRAM, and can be used as non-volatile transistors by replacing gate dielectrics with ferroelectric. The two stable remanent polarization states of the current ferroelectric device are allowed to adjust the transistor Vth value even when the supply voltage is removed [12]. Thus, binary or multi-state is encoded at the threshold voltage of the transistor. The writing of memory cells can be done by changing the polarization state of the erroelectric material and applying pulses to the gate of the transistor affecting the threshold voltage as seen in Fig. 3.

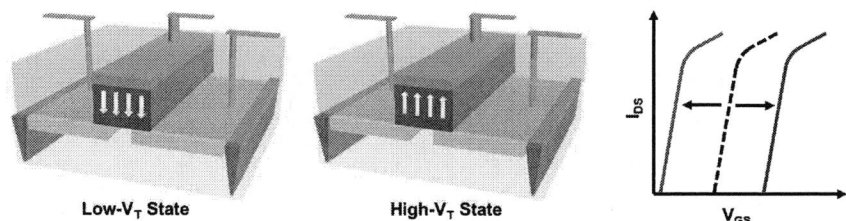

Figure 3. Schematic of FeFET with low V_T state (left) & high V_T state (middle) and transfer curves with different V_T states (right)

Bio-inspired Electronic skin for artificial nerve system

In recent, technology advances in neuromorphic is emerging. Neuromorphic is an electronic device that mimics the operation principle of a human brain. Since the conventional computing system has a separated component of the central processor unit and memory, it has a limitation on speed and the amount of storage. However, our brain can function as both computations and memorize in parallel. In addition to this, the system consisted of a sensor and neuromorphic device can be an artificial nerve [13]. In this approach, we currently working on a ferroelectric hafnia based artificial nerve system. Ferroelectric is well-known material with an interesting behavior that can retain its electrical dipole characteristics even after removing the external field. Also, all ferroelectric material has pyroelectric and piezoelectric properties that respond to heat and pressure respectively. In comparison with lead-zirconium-titanium-oxide (PZT) which is widely used conventional ferroelectric material, hafnia has no lead component and was able to use at nm scale. We hope that the approaches in our laboratory can be widely adopted in various industrial fields such as electronic skin for humanoid robots, health-care monitoring systems, and advanced prosthetic devices.

Summary

In our research, we have developed hafnia-based ferroelectric materials, processes and devices. The different characteristics of the hafnia ferroelectric materials are required for NAND Flash, processing-in-memory, DRAM, sensor, and thin-film transistors for display. We are making it the best fit for each application by the hafnia ferroelectric process development, gate stack and device design. In this presentation, we will examine the issues and requirements of each of these devices, see how we are solving these issues, summarize the remaining issues to date, and discuss the possibility to further improve speed, reliability and the memory window.

Acknowledgments

This work was supported by the BK21 plus program through the National Research Foundation (NRF) funded by the Ministry of Education of Korea. This work was supported in part by the Ministry of Trade, Industry and Energy, Korea, under Project Nos. 10067789, and in part by the Korea Semiconductor Research Consortium Support Program for the development of the future semiconductor device. This work was also supported by Grant Nos. NRF-2019M3F3A1A02071969 and NRF-2019M3F3A1A02071966.

References

[1]   T. Böscke, J. Müller, D. Bräuhaus, U. Schröder, and U. Böttger, "Ferroelectricity in hafnium oxide thin films," *Applied Physics Letters,* vol. 99, p. 102903, 2011.

[2]   M. H. Park, Y. H. Lee, H. J. Kim, Y. J. Kim, T. Moon, K. D. Kim, *et al.*, "Ferroelectricity and antiferroelectricity of doped thin HfO2-based films," *Advanced Materials,* vol. 27, pp. 1811-1831, 2015.

[3]   M. H. Park, Y. H. Lee, T. Mikolajick, U. Schroeder, and C. S. Hwang, "Review and perspective on ferroelectric HfO 2-based thin films for memory applications," *MRS Communications,* vol. 8, pp. 795-808, 2018.

[4]   Z. Fan, J. Chen, and J. Wang, "Ferroelectric HfO2-based materials for next-generation ferroelectric memories," *Journal of Advanced Dielectrics,* vol. 6, p. 1630003, 2016.

[5]   J. Müller, E. Yurchuk, T. Schlösser, J. Paul, R. Hoffmann, S. Müller, *et al.,* "Ferroelectricity in HfO 2 enables nonvolatile data storage in 28 nm HKMG," in *2012 Symposium on VLSI Technology (VLSIT),* 2012, pp. 25-26.

[6]   T. Mikolajick, U. Schroeder, and S. Slesazeck, "Hafnium oxide based ferroelectric devices for memories and beyond," in *2018 International Symposium on VLSI Technology, Systems and Application (VLSI-TSA),* 2018, pp. 1-2.

[7]   M. Hoffmann, U. Schroeder, T. Schenk, T. Shimizu, H. Funakubo, O. Sakata, *et al.,* "Stabilizing the ferroelectric phase in doped hafnium oxide," *Journal of Applied Physics,* vol. 118, p. 072006, 2015.

[8]   M. Hyuk Park, H. Joon Kim, Y. Jin Kim, W. Lee, H. Kyeom Kim, and C. Seong Hwang, "Effect of forming gas annealing on the ferroelectric properties of Hf0. 5Zr0. 5O2 thin films with and without Pt electrodes," *Applied Physics Letters,* vol. 102, p. 112914, 2013.

[9]   Y. Long et al., "A Ferroelectric FET-Based Processing-in-Memory Architecture for DNN Acceleration", *IEEE Journal on Exploratory Solid-State Computational Devices and Circuits,*, vol. 5, p. 113, 2019.

[10]   D. Das et al., "high-k HfZrO ferroelectric insulator by utilizing high pressure anneal", *IEEE Elect. Dev. Lett., in press,* 2020.

[11]   Kai Ni et al., "A circuit compatible accurate compact model for ferroelectric FETs", *Symposiumon VLSI Tech.,* p.131, 2018.

[12]   Kai Ni et al., "Critical Role of Interlayer in $Hf_{0.5}Zr_{0.5}O_2$ Ferroelectric FET Nonvolatile Memory Performance", *IEEE Trans. Elect. Dev.,* Vol. 65, No. 6, p.2461, 2018.

[13]   S. Jeon et al., "Flexible Multimodal Sensors for Electronic Skin: Principle, Materials, Device, Array Architecture, and Data Acquisition Method", *Proceeding of IEEE.,* Vol. 107, No. 10, p.2065, 2019.

P3HT:ZnS based photovoltaic devices with enhanced performance assisted by oxidised carbon nanotubes

Chaohui Wei[a, d], Matthew T. Bishop[a,b], Yifei Wang[b], Fei Gao[b], Chenxu Wang[b], and George Z. Chen[b,c*]

[a] International Doctoral Innovation Centre, Faculty of Science and Engineering, University of Nottingham Ningbo China, Ningbo 315100, P. R. China

[b] Department of Chemical and Environmental Engineering, Faculty of Science and Engineering, The University of Nottingham Ningbo China, Ningbo 315100, People's Republic of China.

[c] Department of Chemical and Environmental Engineering, and Advanced Materials Research Division, Faculty of Engineering, The University of Nottingham, Nottingham NG7 2RD, United Kingdom

[d] College of Energy, Soochow Institute for Energy and Materials InnovationS (SIEMIS), Jiangsu Provincial Key Laboratory for Advanced Carbon Materials and Wearable Energy Technologies, Soochow University, Suzhou 215006, P. R. China

Email: george.chen@nottingham.ac.uk

Photovoltaic devices based of P3HT:ZnS bulk heterojunctions currently produce relatively low efficiencies, limiting the scope these devices have for further development. This work aims to produce a novel photoactive layer, by incorporating oxidised carbon nanotubes (CNTs) into a P3HT:ZnS bulk heterojunction, to investigate the potential CNTs in improving these devices. Furthermore, the active layers of these photovoltaic devices were created using a single-source precursor, thereby limiting potential barriers for scaling up this process to industrial scale. It was found that the CNTs that had been only mildly oxidised (LowOx-CNTs) were longer in length compared to the higher oxidised CNTs (HighOx-CNTs). This allows the HighOx-CNTs to disperse more evenly throughout the active layer, due to the increased oxygen functional groups on the surface of the CNTs, although this does reduce their comparative conductivity. Due to the reduced conductivity of the HighOx-CNTs, 5 and 10 wt% of CNTs performed at lower levels than that of the LowOx-CNTs, with the most notable difference at 5 wt% (0.0064 % and 0.0096 %, respectively). However, LowOx-CNT devices resulted in lower fill factors (FFs), likely due to increased recombination, which can be linked to the increase path length from the increased size of LowOx-CNTs. As the load of the CNTs increased, the FFs for both devices decreased and the Jsc values began to plateau at roughly 0.15 mA/cm$_2$. As the Jsc plateaus, the beneficial conductive properties of LowOx-CNTs was overcome by the detrimental recombination mechanisms introduced by the longer path length. This caused the HighOx-CNTs to outperform the LowOx-CNTs after 15 wt% doping with a power conversion efficiency of 0.0122 %. This increases the efficiency of the undoped ZnS device by more than 7 times, which can be further increased with the incorporation of electron and hole transporting layers.

Introduction

Traditional silicon photovoltaic devices harvest approximately 33 % of the energy from solar radiation, with the remaining energy being lost through either heat emission, unavoidable thermodynamic loses or by simply not being absorbed in the first place (1, 2). This creates a theoretical maximum that can only be overcome by the next generation of photovoltaic devices (2). In this work poly(3-hexylthiophene-2,5-diyl) (P3HT) and Zinc sulphide (ZnS) were used as the electron-donating and electron-accepting materials, respectively. P3HT was selected as it is a well research and utilised electron-donating component of many solar cells (3, 4). ZnS was selected due to its low associated toxicity, an important factor often overlooked when selecting photovoltaic materials (5, 6). To simplify the fabrication of these photovoltaic materials, a single-step deposition method was used with a single-source precursor (6-8). A zinc ethyl xanthate (ZnXan) precursor was selected due to the low temperatures required for decomposition to ZnS, causing little damage to the polymeric component, allowing for a single step deposition of a hybrid bulk heterojunction photovoltaic devices (6, 9-12). However, photovoltaic devices incorporating these two materials currently produce relatively low efficiencies, which needs to be overcome for these devices to see any commercial success.

Carbon nanotubes (CNTs) appear as a cylindrical shell of 'rolled up' graphitic sheets that possess some useful properties, including a large surface area, high mechanical strength, unique electrical/thermal conductivity, and high chemical stability (13-15). Therefore, CNTs have been noted to be an excellent candidate as a dopant in a wide range of applications, such as catalysts (16), sensors (17), energy conversion (18) and energy storage (19, 20). By incorporating CNTs into the device, charge separation and transfer of charge carriers to the electrodes, before they recombine, can be enhanced (21). CNTs can combine with the π-electrons of conjugated polymers to form additional photovoltaic devices (22). This means CNTs act as electron acceptors but also improve the dissociation of excitons, by providing high electron mobility, allowing for the increase in the generated photocurrent (18, 21, 22). However, this application has never been utilised in a photovoltaic device containing both organic and inorganic components deposited via a single-source precursor. Therefore, the novel application of oxidised CNTs was investigated and compared to investigate their feasibility of improving the performance of P3HT hybrid bulk heterojunction photovoltaic devices.

Material and method

Chemicals purchase source

The precursors for the formation of ZnS (ZnCl$_2$ and Potassium ethyl xanthate) were purchased from China Reagent, with a purity of 95 %. Pyridine and 1,2-Dichlorobenzene were also purchased from China reagent, with a purity of 99 %. Regioregular P3HT was purchased from Sigma-Aldrich, with a 99.995 % purity.

The purchased carbon nanotubes used in this research are multi-wall carbon tubes from Cnano Technology Limited cooperation, Beijing, China. The type of CNTs used was multi-walled CNTs (Flotube 9000/9001) with diameter range of 10~15 nm and purity of 95 wt%. Sulfuric acid (H_2SO_4, 95 wt%), nitric acid (HNO_3, 65 wt%), hydrogen peroxide (H_2O_2, 30 wt%). Both CNTs used in this work were synthesized using methods previously reported by Wei et al. 2019 (15).

Synthesis of zinc ethyl xanthate precursors

The method for synthesizing the ZnXan precursor was similar that that previously reported by Bishop et al (6, 8). 0.7 g of zinc chloride dissolved in 20 ml of de-ionized water. This solution was added to 3.5 g of potassium ethyl xanthate dissolved in 20 ml of di-ionized water and stirred for 24 hours revealing a white precipitate (ZnXan). This precipitate was filtered, washed with de-ionized water and dried at 50 °C.

^{1}H NMR (300 MHz, $CDCl_3$): = 4.59 (q, 4H), 1.49 (t, 6H) ppm.

Calc. for $C_6H_{10}O_2S_4Zn$ (%): C 23.4, S 41.7; found: C 23.6, S 41.4.

Fabrication of photovoltaic cell experiments

The mass of P3HT and zinc ethyl xanthate used was kept constant at 5.0 mg and 62.35 mg respectively, in order to obtain a one to one weight ratio of P3HT:ZnS. These were dissolved into a solution of 5.0 ml 1,2-dichlorobenzene (DCB) with 1.0 ml pyridine. 5, 10, 15 and 20 wt% ratios of the CNTs were used and added into the precursor solutions. To ensure effective dispersion, the solutions were sonicated for 5 minutes prior to deposition. After sonication these solutions were then pipetted onto Indium Tin Oxide (ITO) coated glass substrates and spin coated at 1000 rpm for 30 seconds. Devices were then heated to 160 °C for 30 minutes, to decompose the xanthate precursors into ZnS.

Additional Characterization Techniques

A Zeiss ΣIGMA Scanning Electron Microscopy (SEM) operating at 5.0 kV was used for knowing the morphology, size of the deposited films.

Optical characteristics of the samples were measured at room temperature using a Cary 5000 UV-vis absorption spectrometer at a rate of 400 nm/s, between 350 – 750 nm. One cycle was taken for each sample.

Current-voltage measurements were made using a CHI600E electrochemical workstation under illumination at AM1.5 G (100 mW cm^{-2}) light intensity.

NMR resonance 1HNMR spectra were recorded on a Bruker Ultrashield 300, using $CDCl_3$ solutions at 300 MHz to the singlet of $CDCl_3$ at 7.26 ppm and Elemental analysis was carried out using Perkin Elmer Series II CHNS/O Analyzer.

Results and Discussion

Two different types of oxidised CNTs were selected (see Figure 1):

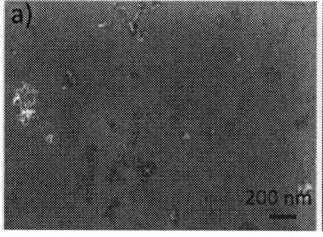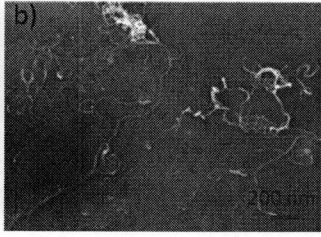

Figure 1. SEM images showing the two different oxidised CNTs used as additives in this study. a) shows CNTs with a high level of oxidation (HighOx-CNTs) and b) shows CNTs with a low level of oxidation (LowOx-CNTs).

- High level of oxidation (HighOx-CNTs);

o These CNTs are shorter, with a narrow size distribution, and contain more oxygen based functional groups (as shown in Figure 1a)) [15]. This will help maintain suspension in a polar solvent, like pyridine.

o HighOx-CNTs were oxidised using a Conventional Acid-modified treatment of CNTs with a mixed acid of H_2SO_4 (95 wt%) and HNO_3 (65 wt%), at a weight ratio of 3:1 [15].

- Low level of oxidation (LowOx-CNTs);

o These CNTs are longer, with a wide size distribution, and contain less oxygen based functional groups (as displayed in Figure 1b)).

o LowOx-CNTs were oxidised using a previously reported Microwave digested treatment, with a mixture of H_2O_2 and H_2SO_4; commonly referred to as piranha solution [15].

Hybrid bulk heterojunction photovoltaic were then doped at 5, 10, 15 and 20 wt% of LowOx-CNTs and HighOx-CNTs. Initially, Tauc plots (see Figure 2) of the deposited thin films were generated by plotting $(\alpha h \nu)2$ vs $h \nu$ (23), as optical properties of photovoltaic devices are very important.

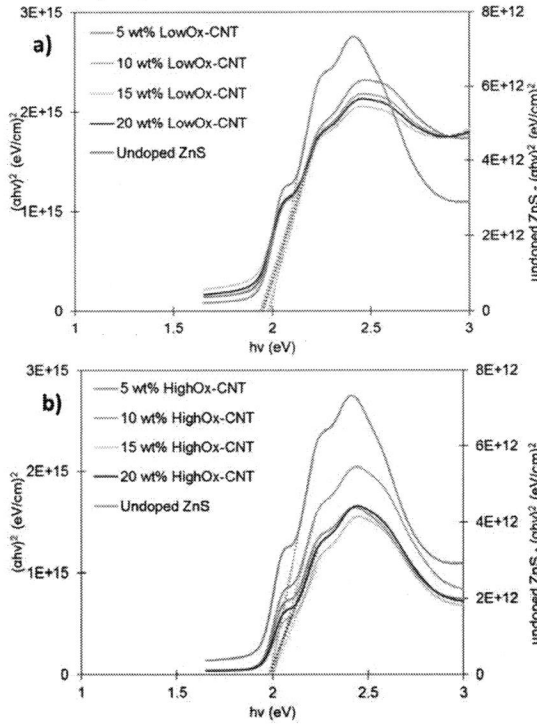

Figure 2. Generated Tauc plots of different photovoltaic devices with a) LowOx-CNTs and b) HighOx-CNTs.

The optical band gap is estimated from Figure 2, by extrapolating straight lines to intercept with the X-axis. It can be seen that all extrapolations intercept the X-axis at roughly the same value, of 2 eV, the band gap value for P3HT (24, 25). It is important to note that no significant shifts in the band gap indicate that the energy level of the valence band maximum (VBM) and conduction band minimum (CBM) have not significantly changed with the incorporation of CNTs. This would suggest that the CNTs do not alter the route for photocurrent generation and that the open circuit voltage would remain relatively unchanged.

To understand how these layers function as a photovoltaic device, current-voltage (I-V) plots were taken for each sample. Figure 3 shows a representative sample, comparing the difference between an undoped and 5 wt% doped photovoltaic devices with two oxidised CNTs. The power conversion efficiency (PCE) was calculated from the experimentally collected values of open circuit voltage (Voc), short-circuit current density (Jsc) and fill factor (FF). These key photovoltaic parameters are summarized in Table 1.

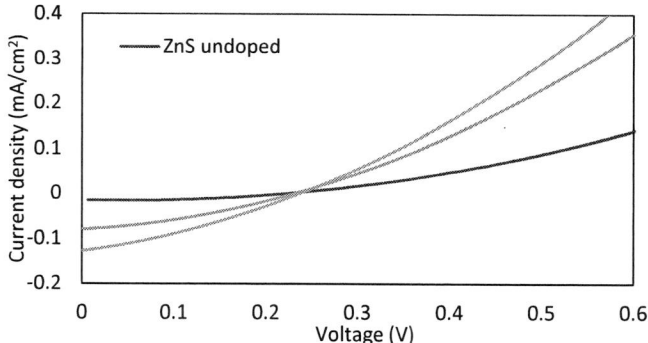

Figure 3. Initial I-V plots collected for undoped and 5 wt% doped photovoltaic devices with two oxidized CNTs, under illumination of 100 mW/cm^2.

Compared to the undoped device, devices doped with 5 wt % of either CNT showed significant increase of Jsc, with the LowOx-CNT producing the highest Jsc. This is reflected in the I-V plots from Figure 3.

TABLE I. Summary of key photovoltaic parameter of samples. Namely the short circuit current (Jsc), open circuit voltage (Voc), the fill factor (FF) and the power conversion efficiency (PCE).

Samples	Jsc (mA/cm^2)	Voc (V)	FF	PCE (%)
Undoped	0.016	0.227	0.47	0.0017
5 wt% LowOx-CNT	0.127	0.238	0.32	0.0096
10 wt% LowOx-CNTs	0.139	0.251	0.3	0.0104
15 wt% LowOx-CNTs	0.15	0.248	0.29	0.0108
20 wt% LowOx-CNTs	0.16	0.231	0.28	0.0105

5 wt% HighOx-CNT	0.08	0.232	0.34	0.0064
10 wt% HighOx-CNTs	0.12	0.238	0.34	0.0098
15 wt% HighOx-CNTs	0.152	0.242	0.33	0.0122
20 wt% HighOx-CNTs	0.156	0.241	0.32	0.0119
Undoped	0.016	0.227	0.47	0.0017
5 wt% LowOx-CNT	0.127	0.238	0.32	0.0096
10 wt% LowOx-CNTs	0.139	0.251	0.3	0.0104
15 wt% LowOx-CNTs	0.15	0.248	0.29	0.0108
20 wt% HighOx-CNTs	0.156	0.241	0.32	0.0119

A continued rise in the value of Jsc was noticed by increasing the CNTs content (**TABLE I.**). This increase in Jsc is related to more favourable transport mechanisms due to the incorporation of CNTs, a phenomena previously noted when doping P3HT:ZnS devices with graphene oxide (5). Initially, the LowOx-CNT produced the highest Jsc, due to their higher conductivity of the CNTs (15). However, 15 and 20 wt% samples resulted in relatively similar Jsc values for both CNTs, suggesting a possible threshold level of current in the device, due to resistances between various interfaces in the device. This means that these values could be dramatically improved through various methods, such as energy level optimisation via the introduction of additional electron and hole transporting layers to enhance carrier mobility and current collection (24, 26).

Unlike the Jsc, the Voc remains more or less constant with or without CNT doping. This shows that the donor and acceptor materials within the photovoltaic device also remain relatively unchanged, an observation described in Figure 3. This would suggest that, despite CNTs potential to function as an acceptor material, in this system they play only a minor role in altering exciton dissociation within the photovoltaic device. This means that the CNT additive works primarily as a charge transport medium, which increases the path length of the material.

The fill factor (FF) was calculated to be 0.47 for the undoped sample, and dramatically decreased when doped with either CNT. This suggests that the presence of the CNT did not result in the suppression of the recombination of excitons, as suggested in literature (27). The decrease in FF is likely due to agglomeration of CNTs within regions of the device causing localised recombination. This means that the incorporation of CNTs into these photovoltaic devices results in an increase in both beneficial transport mechanisms and detrimental recombination mechanisms. LowOx-CNTs resulted in a lower FF, suggesting more recombination, potentially due to their increased size. This increase in chain length increases the size of agglomerated nanoparticles clusters, which function as exciton 'traps'. As the amount of CNTs increased within the devices, the FF reduces further, showing increased agglomeration.

PCEs for both devices peaked at 15 wt%, with HighOx-CNTs producing the most efficient device. However, it is important to note that LowOx-CNTs produced the more efficient device at lower loads, as the increase in initial Jsc was more dramatic, this trend can be seen in Figure 4.

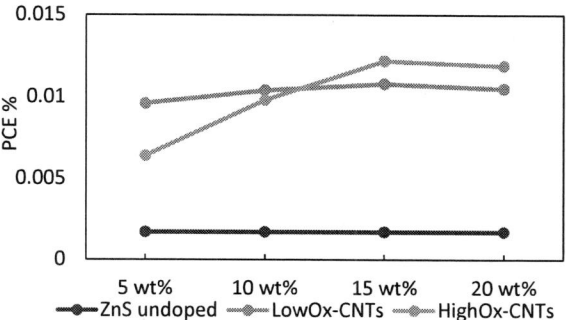

Figure 4. A plot showing trends seen in efficiency as the load of CNT increases. The undoped sample is shown as a reference.

Conclusions

This pilot investigation was undertaken to show the potential benefits that could be gained from incorporating CNTs into P3HT:ZnS based photovoltaic devices to improve their PCEs.

It was shown that after just a 5 wt% addition of either CNT the performance of the device dramatically increased. At lower loads, LowOx-CNTs outperformed the HighOx-CNTs. This can be attributed to the enhanced conductivity of LowOx-CNTs. However, LowOx-CNTs also dramatically reduced the FFs likely due to increased recombination as a result of their larger size. As the loads of the CNTs increased the FFs for both devices decreased, with LowOx-CNTs presenting the lowest FFs. This caused the HighOx-CNTs to outperform the LowOx-CNTs after 15 wt% doping, displaying the highest PCE of 0.0122 %. It was also suggested that this result could be dramatically improved by introducing additional electron and hole transporting layers to enhance carrier mobility and current collection

Regardless, these results show the potential of using CNTs as a dopant in hybrid solar cells, allowing for the improved performance of these devices.

Acknowledgement

This work is supported by Natural Science Foundation of China with a project code 51804172; Ningbo Science and Technology Innovation 2025 Key Project (2018B10029); and the International Doctoral Innovation Centre, Ningbo Education Bureau, Ningbo Science and Technology Bureau and the University of Nottingham.

References

1. Y. W. Su, W. H. Lin, Y. J. Hsu, and K. H. Wei, , *Small*, **10**(22), 4427-4442 (2014).
2. O. E. Semonin, J. M. Luther, and M. C. Beard, *Mater. Today*, **15**(11), 508-515 (2012).
3. R. K. Bhardwaj, H. S. Kushwaha, J. Gaur, T. Upreti, V. Bharti, V. Gupta, N. Chaudhary, G. D. Sharma, K. Banerjee, and S. Chand., *Mater. Lett.*, **89**, 195-197 (2012).
4. U. Mehmood, A. Al-Ahmed, and I. A. Hussein, Renew. *Sust. Energ. Rev.*, **57**, 550-561 (2016).

5. S. Aslam, F. Mustafa, M. A. Ahmad, M. Saleem, M. Idrees, and A. S. Bhatti, *Ceramics Int.*, **44**(1), 402-408 (2018).

6. M. T. Bishop, M. Tomatis, W. Zhang, C. Peng, G. Z. Chen, J. He, and D. Hu, *Sust. Energ & Fuels*, **3**(4), 948-955 (2019).

7. V. Agrawal, K. Jain, L. Arora, and S. Chand, *J. Nanopart. Res.*, **15**(6), 14 (2013).

8. M. Bishop, L. Zhang, A. Dolganov, G. Chen, C. Peng, and D. Hu, *Thin Solid Films*, 137530 (2019).

9. H. C. Leventis, S. P. King, A. Sudlow, M. S. Hill, K. C. Molloy, and S. A. Haque, *Nano Lett.*, **10**(4), 1253-1258 (2010).

10. A. J. MacLachlan, F. T. F. O'Mahony, A. L. Sudlow, M. S. Hill, K. C. Molloy, J. Nelson, and S. A. Haque, *Chemphyschem*, **15**(6), 1019-1023 (2014).

11. P. S. Nair, T. Radhakrishnan, N. Revaprasadu, G. Kolawole, and P. O'Brien, *J. Mater. Chem.*, **12**(9), 2722-2725 (2002).

12. C. Peng, J. Jin, and G. Z. Chen, *Electrochim. Acta*, **53**(2), 525-537 (2007).

13. V. Datsyuk, M. Kalyva, K. Papagelis, J. Parthenios, D. Tasis, A. Siokou, I. Kallitsis, and C. Galiotis, *Carbon*, **46**(6), 833-840 (2008).

14. P.-C. Ma, N. A. Siddiqui, G. Marom, and J.-K. Kim, *Com. Part A: Appl. Sci. and Manufacturing*, **41**(10), 1345-1367, (2010).

15. C. Wei, B. Akinwolemiwa, Q. Wang, L. Guan, L. Xia, D. Hu, B. Tang, L. Yu, and G. Z. Chen, *Adv. Sus. Sys.*, p. 1900065 (2019).

16. W. Li, J. Liu, and C. Yan, *Carbon*, **49**(11), 3463-3470 (2011).

17. Q. Zhao, Z. Gan, and Q. Zhuang, *Electroanalysis*, **14**(23), 1609-1613 (2002).

18. N. A. Nismy, K. Jayawardena, A. Adikaari, and S. R. P. Silva, *Org. Electron.*, **22**, 35-39, (2015).

19. J. H. Chae, X. Zhou, and G. Z. Chen, *green*, **2**(1), 41-54 (2012).

20. C. Wei, B. Akinwolemiwa, L. Yu, D. Hu, and G. Z. Chen, "7 - Polymer Composites with Functionalized Carbon Nanotube and Graphene," in Polymer Composites with Functionalized Nanoparticles, K. Pielichowski and T. M. Majka, Eds.: *Elsevier*, 211-248 (2019).

21. I. Khatri, S. Adhikari, H. R. Aryal, T. Soga, T. Jimbo, and M. Umeno, *Appl. Phys. Lett.*, **94**(9), 3, (2009).

22. S. P. Somani, P. R. Somani, and M. Umeno, *Diamond and Related Mat.*, **17**(4-5), 585-588, (2008).

23. M. A. Buckingham, A. L. Catherall, M. S. Hill, A. L. Johnson, and J. D. Parish, *Cryst. Growth Des.*, **17**(2), 907-912, (2017).

24. R. Kroon, M. Lenes, J. C. Hummelen, P. W. M. Blom, and B. De Boer, *Polym. Rev.*, **48**(3), 531-582 (2008).

25. I. Borazan, Y. Altin, A. Demir, and A. C. Bedeloglu, *J. Polym. Eng.*, **39**(7), 636-641 (2019).

26. Y. S. Kwon, J. Lim, H. J. Yun, Y. H. Kim, and T. Park, *Energy Environ. Sci.*, **7**(4), 1454-1460 (2014).

27. S. Englund, V. Paneta, D. Primetzhofer, Y. Ren, O. Donzel-Gargand, J. K. Larsen, J. Scragg, C. P. Bjorkman., *Thin Solid Films*, **639**, 91-97 (2017).

Author Index

Aoki, T.	185	Ikeda, T.	13, 55
		Inguva, S.	151
Baba, H.	55	Ito, M.	13
Belmonte, A.	205		
Bender, M.	69	Jeon, S.	219
Bermundo, J. P.	29	Jiang, C.	173
Borniquel,J.I.D.A.	205	Jiang, J.	39
Buckley, D.	151		
		Kabir, M. S.	81
Chasin, A.	205	Kar, G. S.	205
Chen, G. Z.	225	Kim, H.	97, 109, 117
Choi, S. Y.	69	Kim, J. B.	69
Choi, Y.	47	Kim, M.	47
Chowdhury, R. R.	81	Kim, S. H.	47
Corsino, D.	29	Kimura, H.	55, 185
		Kljucar, L.	205
Dekkers, H. F. W.	205	Kozuma, M.	185
Delhougne, R.	205	Kunitake, H.	13, 55, 185
Donadio, G. L.	205	Kuo, Y.	39
Fujii, M. N.	3, 29	Lim, J. H.	47
Furuta, M.	89	Lim, R.	69
		Liu, X.	161
Gao, F.	225		
Gity, F.	151	Magari, Y.	89
Godo, H.	55	Manley, R. G.	81, 97, 109, 117, 131,
Goux, L.	205		141
		Matsumoto, N.	13
Hao, M.	69	Matsuzaki, T.	185
Hirschman, K. D.	81, 131, 141	McNulty, D.	151
Huang, M. H.	97, 109, 117	Mitard, J.	205
Hum, M.	131	Miyanaga, R.	3
Hurley, P. K.	151	Miyata, S.	13
		Mizukami, S.	13

Motoyoshi, R.	185	Tanabe, H.	3
Murakawa, T.	55	Tanaka, J.	3
		Tokei, Z.	205
Nag, M.	205	Tsai, Y. C.	69
Nathan, A.	173	Tsuda, K.	13, 55
O'Dwyer, C.	151	Uenuma, M.	29
Oh, S.	47	Uraoka, Y.	3, 29
Ohshima, K.	13		
Okamoto, Y.	185	Vaddi, R.	97, 109, 117
Okuno, N.	13	van Setten, M.	205
Onuki, T.	185		
		Wang, C.	225
Packard, G.	131, 141	Wang, S. W.	69
Parbrook, P.	151	Wang, Y.	225
Park, J. S.	47	Wang, Z.	69
Park, J. W.	69	Wei, C.	225
Park, J.	161	Wilson, C.	205
Pourtois, G.	205		
		Yakubo, Y.	13
Rassoul, N.	205	Yamane, Y.	185
Rosenfeld, A.	131, 141	Yamazaki, S.	13, 55, 185
		Yim, D. K.	69
Sasagawa, S.	13, 55, 185	Ytterdal, T.	161
Sawai, H.	55	Yu, S. M.	47
Shi, Y.	117	Yuichi, Y.	13
Shur, M.	161		
Steudel, S.	205	Zhang, Y.	161
Suzuki, A.	13	Zhao, L.	69
		Zhu, B.	97, 109, 117
T. Bishop, M.	225	Zubialevich, V. Z.	151
Takahashi, T.	3		
Takechi, K.	3		